GEOTECHNICAL ASPECTS
IN THE SITING AND DESIGN
OF NUCLEAR INSTALLATIONS

The following States are Members of the International Atomic Energy Agency:

AFGHANISTAN	GEORGIA	OMAN
ALBANIA	GERMANY	PAKISTAN
ALGERIA	GHANA	PALAU
ANGOLA	GREECE	PANAMA
ANTIGUA AND BARBUDA	GRENADA	PAPUA NEW GUINEA
ARGENTINA	GUATEMALA	PARAGUAY
ARMENIA	GUINEA	PERU
AUSTRALIA	GUYANA	PHILIPPINES
AUSTRIA	HAITI	POLAND
AZERBAIJAN	HOLY SEE	PORTUGAL
BAHAMAS, THE	HONDURAS	QATAR
BAHRAIN	HUNGARY	REPUBLIC OF MOLDOVA
BANGLADESH	ICELAND	ROMANIA
BARBADOS	INDIA	RUSSIAN FEDERATION
BELARUS	INDONESIA	RWANDA
BELGIUM	IRAN, ISLAMIC REPUBLIC OF	SAINT KITTS AND NEVIS
BELIZE	IRAQ	SAINT LUCIA
BENIN	IRELAND	SAINT VINCENT AND
BOLIVIA, PLURINATIONAL	ISRAEL	THE GRENADINES
STATE OF	ITALY	SAMOA
BOSNIA AND HERZEGOVINA	JAMAICA	SAN MARINO
BOTSWANA	JAPAN	SAUDI ARABIA
BRAZIL	JORDAN	SENEGAL
BRUNEI DARUSSALAM	KAZAKHSTAN	SERBIA
BULGARIA	KENYA	SEYCHELLES
BURKINA FASO	KOREA, REPUBLIC OF	SIERRA LEONE
BURUNDI	KUWAIT	SINGAPORE
CABO VERDE	KYRGYZSTAN	SLOVAKIA
CAMBODIA	LAO PEOPLE'S DEMOCRATIC	SLOVENIA
CAMEROON	REPUBLIC	SOMALIA
CANADA	LATVIA	SOUTH AFRICA
CENTRAL AFRICAN	LEBANON	SPAIN
REPUBLIC	LESOTHO	SRI LANKA
CHAD	LIBERIA	SUDAN
CHILE	LIBYA	SWEDEN
CHINA	LIECHTENSTEIN	SWITZERLAND
COLOMBIA	LITHUANIA	SYRIAN ARAB REPUBLIC
COMOROS	LUXEMBOURG	TAJIKISTAN
CONGO	MADAGASCAR	THAILAND
COOK ISLANDS	MALAWI	TOGO
COSTA RICA	MALAYSIA	TONGA
CÔTE D'IVOIRE	MALDIVES	TRINIDAD AND TOBAGO
CROATIA	MALI	TUNISIA
CUBA	MALTA	TÜRKİYE
CYPRUS	MARSHALL ISLANDS	TURKMENISTAN
CZECH REPUBLIC	MAURITANIA	UGANDA
DEMOCRATIC REPUBLIC	MAURITIUS	UKRAINE
OF THE CONGO	MEXICO	UNITED ARAB EMIRATES
DENMARK	MONACO	UNITED KINGDOM OF
DJIBOUTI	MONGOLIA	GREAT BRITAIN AND
DOMINICA	MONTENEGRO	NORTHERN IRELAND
DOMINICAN REPUBLIC	MOROCCO	UNITED REPUBLIC OF TANZANIA
ECUADOR	MOZAMBIQUE	UNITED STATES OF AMERICA
EGYPT	MYANMAR	URUGUAY
EL SALVADOR	NAMIBIA	UZBEKISTAN
ERITREA	NEPAL	VANUATU
ESTONIA	NETHERLANDS,	VENEZUELA, BOLIVARIAN
ESWATINI	KINGDOM OF THE	REPUBLIC OF
ETHIOPIA	NEW ZEALAND	VIET NAM
FIJI	NICARAGUA	YEMEN
FINLAND	NIGER	ZAMBIA
FRANCE	NIGERIA	ZIMBABWE
GABON	NORTH MACEDONIA	
GAMBIA, THE	NORWAY	

The Agency's Statute was approved on 23 October 1956 by the Conference on the Statute of the IAEA held at United Nations Headquarters, New York; it entered into force on 29 July 1957. The Headquarters of the Agency are situated in Vienna. Its principal objective is "to accelerate and enlarge the contribution of atomic energy to peace, health and prosperity throughout the world".

IAEA SAFETY STANDARDS SERIES No. SSG-93

GEOTECHNICAL ASPECTS IN THE SITING AND DESIGN OF NUCLEAR INSTALLATIONS

SPECIFIC SAFETY GUIDE

INTERNATIONAL ATOMIC ENERGY AGENCY
VIENNA, 2026

IAEA Library Cataloguing in Publication Data

Names: International Atomic Energy Agency.
Title: Geotechnical aspects in the siting and design of nuclear installations / International Atomic Energy Agency.
Description: Vienna : International Atomic Energy Agency, 2026. | Series: Safety standards series, ISSN 1020-525X ; no. SSG-93 | Includes bibliographical references.
Identifiers: IAEAL 26-01807 | ISBN 978-92-0-116725-5 (paperback : alk. paper) | ISBN 978-92-0-116825-2 (pdf) | ISBN 978-92-0-116925-9 (epub)
Subjects: LCSH: Nuclear facilities — Design and construction. | Nuclear facilities — Safety measures. | Geotechnical engineering.
Classification: UDC 621.039.58 | STI/PUB/2123

FOREWORD

by Rafael Mariano Grossi
Director General

The IAEA's Statute authorizes it to "establish...standards of safety for protection of health and minimization of danger to life and property". These are standards that the IAEA must apply to its own operations, and that States can apply through their national regulations.

The IAEA started its safety standards programme in 1958 and there have been many developments since. As Director General, I am committed to ensuring that the IAEA maintains and improves upon this integrated, comprehensive and consistent set of up to date, user friendly and fit for purpose safety standards of high quality. Their proper application in the use of nuclear science and technology should offer a high level of protection for people and the environment across the world and provide the confidence necessary to allow for the ongoing use of nuclear technology for the benefit of all.

Safety is a national responsibility underpinned by a number of international conventions. The IAEA safety standards form a basis for these legal instruments and serve as a global reference to help parties meet their obligations. While safety standards are not legally binding on Member States, they are widely applied. They have become an indispensable reference point and a common denominator for the vast majority of Member States that have adopted these standards for use in national regulations to enhance safety in nuclear power generation, research reactors and fuel cycle facilities as well as in nuclear applications in medicine, industry, agriculture and research.

The IAEA safety standards are based on the practical experience of its Member States and produced through international consensus. The involvement of the members of the Safety Standards Committees, the Nuclear Security Guidance Committee and the Commission on Safety Standards is particularly important, and I am grateful to all those who contribute their knowledge and expertise to this endeavour.

The IAEA also uses these safety standards when it assists Member States through its review missions and advisory services. This helps Member States in the application of the standards and enables valuable experience and insight to be shared. Feedback from these missions and services, and lessons identified from events and experience in the use and application of the safety standards, are taken into account during their periodic revision.

I believe the IAEA safety standards and their application make an invaluable contribution to ensuring a high level of safety in the use of nuclear technology. I encourage all Member States to promote and apply these standards, and to work with the IAEA to uphold their quality now and in the future.

THE IAEA SAFETY STANDARDS

BACKGROUND

Radioactivity is a natural phenomenon and natural sources of radiation are features of the environment. Radiation and radioactive substances have many beneficial applications, ranging from power generation to uses in medicine, industry and agriculture. The radiation risks to workers and the public and to the environment that may arise from these applications have to be assessed and, if necessary, controlled.

Activities such as the medical uses of radiation, the operation of nuclear installations, the production, transport and use of radioactive material, and the management of radioactive waste must therefore be subject to standards of safety.

Regulating safety is a national responsibility. However, radiation risks may transcend national borders, and international cooperation serves to promote and enhance safety globally by exchanging experience and by improving capabilities to control hazards, to prevent accidents, to respond to emergencies and to mitigate any harmful consequences.

States have an obligation of diligence and duty of care, and are expected to fulfil their national and international undertakings and obligations.

International safety standards provide support for States in meeting their obligations under general principles of international law, such as those relating to environmental protection. International safety standards also promote and assure confidence in safety and facilitate international commerce and trade.

A global nuclear safety regime is in place and is being continuously improved. IAEA safety standards, which support the implementation of binding international instruments and national safety infrastructures, are a cornerstone of this global regime. The IAEA safety standards constitute a useful tool for contracting parties to assess their performance under these international conventions.

THE IAEA SAFETY STANDARDS

The status of the IAEA safety standards derives from the IAEA's Statute, which authorizes the IAEA to establish or adopt, in consultation and, where appropriate, in collaboration with the competent organs of the United Nations and with the specialized agencies concerned, standards of safety for protection of health and minimization of danger to life and property, and to provide for their application.

With a view to ensuring the protection of people and the environment from harmful effects of ionizing radiation, the IAEA safety standards establish fundamental safety principles, requirements and measures to control the radiation exposure of people and the release of radioactive material to the environment, to restrict the likelihood of events that might lead to a loss of control over a nuclear reactor core, nuclear chain reaction, radioactive source or any other source of radiation, and to mitigate the consequences of such events if they were to occur. The standards apply to facilities and activities that give rise to radiation risks, including nuclear installations, the use of radiation and radioactive sources, the transport of radioactive material and the management of radioactive waste.

Safety measures and security measures[1] have in common the aim of protecting human life and health and the environment. Safety measures and security measures must be designed and implemented in an integrated manner so that security measures do not compromise safety and safety measures do not compromise security.

The IAEA safety standards reflect an international consensus on what constitutes a high level of safety for protecting people and the environment from harmful effects of ionizing radiation. They are issued in the IAEA Safety Standards Series, which has three categories (see Fig. 1).

Safety Fundamentals

Safety Fundamentals present the fundamental safety objective and principles of protection and safety, and provide the basis for the Safety Requirements. The principles are expressed as 'must' statements.

Safety Requirements

Safety Requirements are governed by the objective and principles of the Safety Fundamentals. They establish the requirements to be met to ensure the protection of people and the environment, both now and in the future. The format and style of the Safety Requirements facilitate their use for the establishment of a national regulatory framework. Requirements are presented as 'overarching' requirements[2] in bold, followed by a number of associated requirements; all are equally important and are expressed as 'shall' statements.

Safety Guides

Safety Guides provide recommendations on how to comply with the Safety Requirements, indicating an international consensus that it is necessary to take the

---

[1]   See also publications issued in the IAEA Nuclear Security Series.

[2]   The IAEA Regulations for the Safe Transport of Radioactive Material do not include overarching requirements.

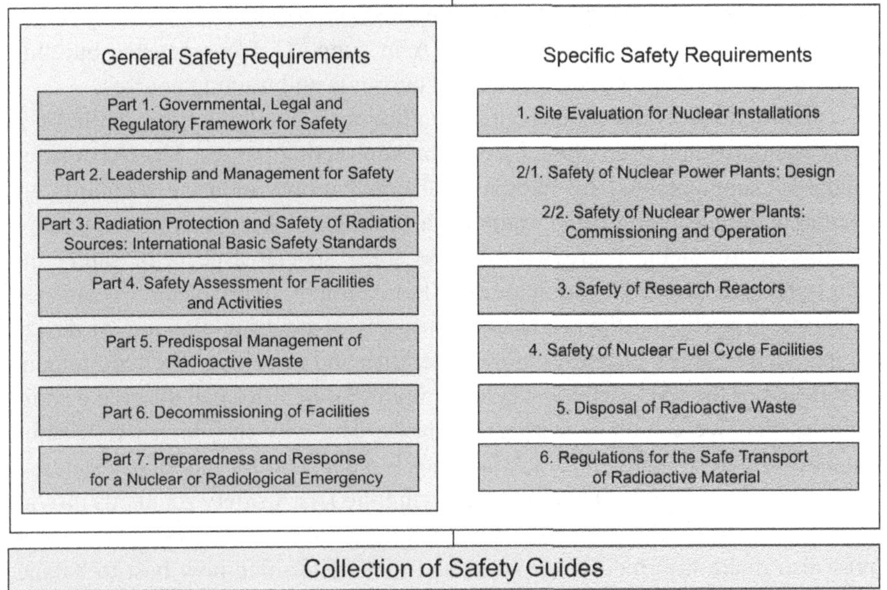

FIG. 1. The long term structure of the IAEA Safety Standards Series.

measures recommended (or alternative measures that achieve the same level of protection). Safety Guides present international good practices and, increasingly, best practices. The recommendations provided in Safety Guides are expressed as 'should' statements.

APPLICATION OF THE IAEA SAFETY STANDARDS

The principal users of safety standards in IAEA Member States are regulatory bodies and other relevant national authorities. The IAEA safety standards are also used by co-sponsoring organizations and by many organizations that design, construct and operate nuclear facilities, as well as organizations involved in the use of radiation and radioactive sources.

The IAEA safety standards are applicable, as relevant, throughout the entire lifetime of all facilities and activities — existing and new — utilized for peaceful purposes and to protective actions to reduce existing radiation risks. They can be used by States as a reference for their national regulations in respect of facilities and activities.

The IAEA's Statute makes the safety standards binding on the IAEA in relation to its own operations and also on States in relation to IAEA assisted operations.

The IAEA safety standards also form the basis for the IAEA's safety review services, and they are used by the IAEA in support of competence building, including the development of educational curricula and training courses.

International conventions contain requirements similar to those in the IAEA safety standards and make them binding on contracting parties. The IAEA safety standards, supplemented by international conventions, industry standards and detailed national requirements, establish a consistent basis for protecting people and the environment. There will also be some special aspects of safety that need to be assessed at the national level. For example, many of the IAEA safety standards, in particular those addressing aspects of safety in planning or design, are intended to apply primarily to new facilities and activities. The requirements established in the IAEA safety standards might not be fully met at some existing facilities that were built to earlier standards. The way in which IAEA safety standards are to be applied to such facilities is a decision for individual States.

The scientific considerations underlying the IAEA safety standards provide an objective basis for decisions concerning safety; however, decision makers must also make informed judgements and must determine how best to balance the benefits of an action or an activity against the associated radiation risks and any other detrimental impacts to which it gives rise.

DEVELOPMENT PROCESS FOR THE IAEA SAFETY STANDARDS

The preparation and review of the safety standards involves the IAEA Secretariat and five Safety Standards Committees, for emergency preparedness and response (EPReSC) (as of 2016), nuclear safety (NUSSC), radiation safety (RASSC), the safety of radioactive waste (WASSC) and the safe transport of radioactive material (TRANSSC), and a Commission on Safety Standards (CSS) which oversees the IAEA safety standards programme (see Fig. 2).

All IAEA Member States may nominate experts for the Safety Standards Committees and may provide comments on draft standards. The membership of the Commission on Safety Standards is appointed by the Director General and includes senior governmental officials having responsibility for establishing national standards.

A management system has been established for the processes of planning, developing, reviewing, revising and establishing the IAEA safety standards. It articulates the mandate of the IAEA, the vision for the future application of the safety standards, policies and strategies, and corresponding functions and responsibilities.

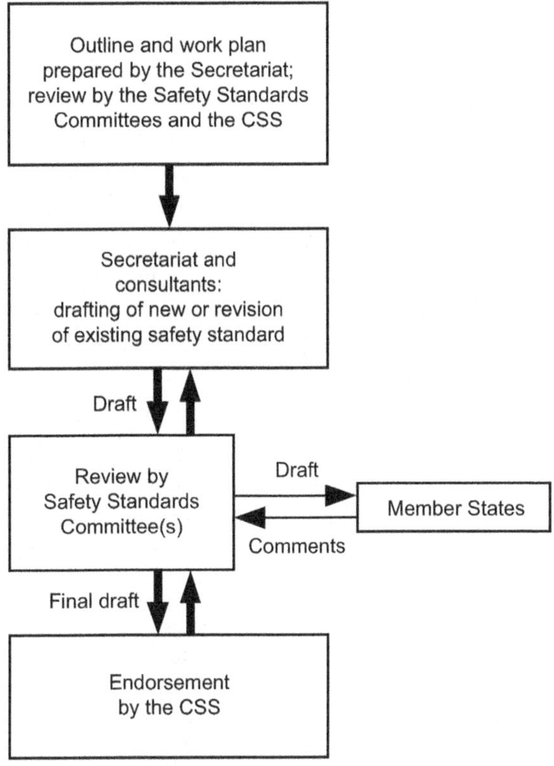

FIG. 2. The process for developing a new safety standard or revising an existing standard.

INTERACTION WITH OTHER INTERNATIONAL ORGANIZATIONS

The findings of the United Nations Scientific Committee on the Effects of Atomic Radiation (UNSCEAR) and the recommendations of international expert bodies, notably the International Commission on Radiological Protection (ICRP), are taken into account in developing the IAEA safety standards. Some safety standards are developed in cooperation with other bodies in the United Nations system or other specialized agencies, including the Food and Agriculture Organization of the United Nations, the United Nations Environment Programme, the International Labour Organization, the OECD Nuclear Energy Agency, the Pan American Health Organization and the World Health Organization.

INTERPRETATION OF THE TEXT

Safety related terms are to be understood as they appear in the IAEA Nuclear Safety and Security Glossary (see https://www.iaea.org/resources/publications/iaea-nuclear-safety-and-security-glossary). Otherwise, words are used with the spellings and meanings assigned to them in the latest edition of The Concise Oxford Dictionary. For Safety Guides, the English version of the text is the authoritative version.

The background and context of each standard in the IAEA Safety Standards Series and its objective, scope and structure are explained in Section 1, Introduction, of each publication.

Material for which there is no appropriate place in the body text (e.g. material that is subsidiary to or separate from the body text, is included in support of statements in the body text, or describes methods of calculation, procedures or limits and conditions) may be presented in appendices or annexes.

An appendix, if included, is considered to form an integral part of the safety standard. Material in an appendix has the same status as the body text, and the IAEA assumes authorship of it. Annexes and footnotes to the main text, if included, are used to provide practical examples or additional information or explanation. Annexes and footnotes are not integral parts of the main text. Annex material published by the IAEA is not necessarily issued under its authorship; material under other authorship may be presented in annexes to the safety standards. Extraneous material presented in annexes is excerpted and adapted as necessary to be generally useful.

CONTENTS

1. INTRODUCTION

BACKGROUND

1.1. Requirements for site evaluation for nuclear installations are established in IAEA Safety Standards Series No. SSR-1, Site Evaluation for Nuclear Installations [1]. This Safety Guide provides recommendations on geotechnical characteristics and the evaluation of geotechnical hazards as part of such a site evaluation.

1.2. Seismic aspects also play an important role in this field, and relevant recommendations are provided in IAEA Safety Standards Series No. SSG-9 (Rev. 1), Seismic Hazards in Site Evaluation for Nuclear Installations [2].

1.3. This Safety Guide supersedes IAEA Safety Standards Series No. NS-G-3.6, Geotechnical Aspects of Site Evaluation and Foundations for Nuclear Power Plants.[1] The revision ensures consistency with the requirements established in SSR-1 [1], while incorporating the latest knowledge, experience and lessons learned from significant geotechnical events in Member States. This Safety Guide explicitly expands the scope to include nuclear installations other than nuclear power plants (excluding nuclear waste disposal facilities) and provides recommendations for applying a graded approach to geotechnical site investigations and activities for other types of nuclear installations.

OBJECTIVE

1.4. The objective of this Safety Guide is to provide recommendations on dealing with geotechnical engineering aspects important to the safety of nuclear installations, such as site investigation planning, evaluation of geotechnical hazards, considerations for design and analyses, monitoring of geotechnical parameters, and the application of a graded approach to geotechnical evaluations for nuclear installations other than nuclear power plants. These recommendations are intended to meet the requirements established in SSR-1 [1], in particular Requirements 21 and 22.

---

[1] INTERNATIONAL ATOMIC ENERGY AGENCY, Geotechnical Aspects of Site Evaluation and Foundations for Nuclear Power Plants, IAEA Safety Standards Series No. NS-G-3.6, IAEA, Vienna (2004).

1.5. This Safety Guide is intended for use by operating organizations, licensees and regulatory bodies involved in the licensing of nuclear installations, as well as by the designers and technical support organizations of such installations.

SCOPE

1.6. In this Safety Guide, 'geotechnical aspects' refer to aspects of geotechnical site investigation, evaluation, engineering design and safety assessment relating to the subsurface conditions at nuclear installation sites.

1.7. This Safety Guide provides recommendations on the geotechnical aspects necessary for the establishment of parameters used in the site evaluation and the development of the design basis for nuclear installations. It covers the programme of site investigation to be implemented to obtain an appropriate understanding of the subsurface conditions, which is necessary for determining whether the conditions are suitable for the foundations and for the construction of a nuclear installation. It provides recommendations specific to the characteristics of the geotechnical profiles (foundation ground types) and the parameters suitable for use in performing geotechnical analyses for the design of a nuclear installation. It also addresses the approach to monitoring geotechnical parameters, the application of a graded approach and the application of a management system.

1.8. This Safety Guide provides recommendations on the methods of analysis appropriate for the safety assessment of a site for a nuclear installation, addressing assessments of all external events, particularly the assessment of earthquake effects on the site, including the determination of site specific response spectra and the estimation of the liquefaction potential. This Safety Guide also provides recommendations on the methods of analysis for the safety assessment of the effects of static and dynamic interaction between soil and structures and of the consequences on the soil bearing capacity and on settlements. A more detailed description of methods for the analysis of soil–structure interaction is given in SSG-9 (Rev. 1) [2]. In this Safety Guide, only the site dependent information and the methods of analysis are addressed.

1.9. This Safety Guide also considers foundation works, which are based on geotechnical profiles and parameters, the possible techniques for the improvement of foundation material, and the choice of foundation system appropriate for the soil capacity. Earth structures, natural slopes and buried structures, the safety of which need to be assessed in the site safety assessment, are also considered. The

Safety Guide provides recommendations on appropriate methods for the analysis of the behaviour of such structures under static and dynamic loads.

1.10. This Safety Guide also provides recommendations on methodologies for the development of the design basis for nuclear installations. The collected data and interpreted information from site investigations (considering their variability and the analysis methodologies described in this Safety Guide) are appropriate for use in the evaluation of structural response to both design basis and beyond design basis events. The acceptance criteria for the assessment of beyond design basis external events may be relaxed, provided that the criteria are consistent with the provisions for beyond design basis external hazards described in IAEA Safety Standards Series Nos SSG-67, Seismic Design for Nuclear Installations [3], and SSG-68, Design of Nuclear Installations Against External Events Excluding Earthquakes [4]. Furthermore, these evaluations need to consider the potential for cliff edge effects and provide adequate margin to protect the items ultimately necessary to prevent an early radioactive release or a large radioactive release.

1.11. This Safety Guide does not include recommendations specific to subsurface nuclear installations, which necessitate a higher level of effort and greater focus on subsurface exploration, tunnel construction and site specific considerations.

STRUCTURE

1.12. Section 2 provides recommendations on geotechnical site investigation, addressing different stages of the programme, sources of data, special considerations for the investigation of complex subsurface conditions, and site considerations for nuclear installations. Section 3 provides recommendations relating to geotechnical hazards, including undesirable subsurface conditions, natural slopes and liquefaction. Section 4 provides recommendations on considerations for the design and evaluation of dykes and dams, seawalls and retaining walls, foundations, earth structures, buried structures, embedded structures, buried pipes, conduits and tunnels. Section 5 provides recommendations on monitoring geotechnical parameters. Section 6 provides recommendations on applying a graded approach to geotechnical aspects for nuclear installations other than nuclear power plants. Section 7 provides recommendations on the application of a management system, with a focus on quality management for geotechnical investigations, testing, verification, record keeping and monitoring.

2. GEOTECHNICAL SITE CONSIDERATIONS FOR NUCLEAR INSTALLATIONS

GEOTECHNICAL INVESTIGATION PROGRAMME FOR THE SITING OF NUCLEAR INSTALLATIONS

2.1.   Requirement 21 of SSR-1 [1] states that "**The geotechnical characteristics and geological features of subsurface materials shall be investigated, and a soil and rock profile for the site that considers the variability and uncertainty in subsurface materials shall be derived.**"

2.2.   Investigations of the subsurface conditions at potential sites for a nuclear installation should be performed at all stages of the site evaluation process (see paras 2.7–2.26). The purpose of such investigations is to obtain information and basic data on the physical and mechanical properties of the subsurface materials, for use in decision making about the suitability of the site for a nuclear installation, and to ensure the safety of the installation throughout its lifetime.

2.3.   The geotechnical investigation programme should provide the data necessary for an appropriate characterization of the subsurface at each stage of the site evaluation of a nuclear installation. The various methods of investigation (i.e. the use of current and historical documents, geological data, geophysical and geotechnical investigations, and in situ and laboratory testing) are typically applicable to all stages of the site evaluation process but vary from stage to stage, as necessary. In general, the investigations should become more detailed in character when approaching the later stages of the investigation programme. Furthermore, some analysis specific considerations may apply only to data sets used as input data in soil and rock characterization and analysis.

2.4.   The long term impact of investigative drilling on the geological environment and aquifers should be considered. Relevant precautions should be taken to eliminate any long term negative impacts. All boreholes not needed for monitoring purposes (see Section 5) should be filled and sealed with suitable materials.

2.5.   Generally, data relating to geophysical, geological, geotechnical and engineering information should be collected for use in safety evaluations or analyses. The data are typically grouped as follows:

(a)   Composition of the subsurface (rock and soil types);
(b)   Characterization of the subsurface (in terms of physical, chemical, geomechanical and filtration properties), including applicable classifications (e.g. those used in engineering geology);
(c)   Spatial information about the continuity, extent and geometrical arrangement of the subsurface materials (e.g. stratigraphy and geometry of geological structures);
(d)   Spatial information and properties of discontinuities and/or other features in the subsurface (e.g. faults, fracture zones, cavities) that could affect the suitability of the site (e.g. in terms of mechanical stability or hydrogeology), including applicable classifications (e.g. those used in engineering geology);
(e)   Hydrogeological, hydrological and hydrochemical information (e.g. groundwater regime, hydrostratigraphical and hydrogeological model, groundwater table, groundwater chemicals, quality of the groundwater, connections between groundwater and surface water);
(f)   Geomorphological information documenting the landforms and terrain features and their interaction with geological processes.

2.6. The results of the investigations should be clearly documented (see para. 2.38) with reference to the particular site conditions (e.g. soil or rock), the stage of the site evaluation process concerned and the verification analysis needed. The detail of this documentation should be sufficient to support the safety justification, evaluations and analyses, as well as to support independent peer reviews and review and assessment by the regulatory body.

Site selection stage

2.7. The purpose of an investigation at the site selection stage should be to determine the preliminary suitability of sites (see para. 2.3 of IAEA Safety Standards Series No. SSG-35, Site Survey and Site Selection for Nuclear Installations [5]). During this stage, geological, geophysical, geochemical, geomorphological, geotechnical, hydrogeological and hydrological aspects are considered, and some regions or areas may be excluded from further consideration. The information on subsurface conditions is usually obtained from historical and current documents (see paras 2.28 and 2.29) and by means of field reconnaissance, including geological, geophysical and geomorphological surveys (see para. 2.30) at this stage. This information is used in the following considerations:

(a)   Unacceptable subsurface conditions. A site with geotechnical characteristics and geological features (as investigated in accordance with Requirement 21 of SSR-1 [1]) that could challenge the safety of a nuclear installation and that cannot be corrected by means of geotechnical treatment or compensated for by design or construction measures is unacceptable. Unacceptable subsurface conditions should be considered to be exclusion criteria. Geotechnical hazards and geological hazards are required to be evaluated (see Requirement 22 of SSR-1 [1]). Furthermore, the potential for geotechnical hazards and geological hazards associated with faulting, ground motion, uneven bedrock movements, flooding, volcanic activity, landslides, permafrost, swelling, erosion processes and migratory sand dunes should be evaluated. The scope and extent of the investigation should be sufficient to estimate the hazard under consideration with a level of confidence that can enable the application of the relevant exclusion criteria.

(b) Classification of the site. The site should be classified for the purpose of seismic site response analysis, using the shear wave velocity profile (see paras 2.42 and 2.43). If such site classification is not yet applicable, the subsurface conditions at a site can be derived from the geological and geotechnical literature, and the site may be classified into one of three main categories: a rock site, a soil site, or a combination of rock and soil site. If applicable, the hardness (soft, medium or hard) of the rock at a rock site should be further classified. If applicable, the stiffness (soft, medium or stiff) of the soil at a soil site should be further classified. However, this approximate classification might not apply to certain sites. For instance, quaternary formations or intensive bedrock fracturing and alteration may introduce complex interfaces and ambiguity in defining the contacts between the different subsurface materials.

(c) Groundwater regime. If there is a lack of detailed data, the hydrogeological literature may allow a preliminary estimation of the presence and level of groundwater, the potential groundwater–surface water interactions and the groundwater regime at this stage. In later stages, further investigations should be performed in accordance with para. 5.26 of SSR-1 [1] and with IAEA Safety Standards Series No. SSG-92, Investigation of Site Characteristics and Evaluation of Radiation Risks to the Public and the Environment in Site Evaluation for Nuclear Installations [6].

(d) Foundation conditions. The type of soil and/or bedrock, its properties, its lateral extent, and the depth to bedrock or load bearing stratum should be determined, as a minimum set of information. This enables the preliminary selection of suitable foundation types.

2.8. On the basis of the information on subsurface conditions, candidate sites can be ranked in accordance with the suitability of foundation works. In addition to the assessment of the potentially unacceptable subsurface conditions (see para. 2.7(a)), inferences can be made about seismic amplification effects, bearing capacity, slope stability, potential settlement and swelling, and soil–structure interactions. After this stage, sites with unacceptable subsurface conditions for which there are no generally practicable engineering solutions should be excluded; sites with acceptable subsurface conditions can be retained for further consideration.

2.9. The geotechnical site investigation programme for a nuclear installation should consider the potential presence of particularly undesirable subsurface conditions, which could have serious implications for the integrity of the foundation of the installation due to ground instability and/or collapse, bedrock block movements, or changes in groundwater conditions. In investigating such undesirable subsurface conditions, the following should be considered:

(a) Potential cavities and susceptibility to ground collapse:
   (i) Underground void spaces, of either natural or artificial origin;
   (ii) Sinkholes and open joints that give rise to hazardous effects of other types, such as piping and seepage;
   (iii) Sinks, sink ponds, caves, cavity zones and caverns;
   (iv) Gas pockets;
   (v) Evidence of solution or karstic phenomena;
   (vi) Sinking streams;
   (vii) Historical ground subsidence;
   (viii) Mines and signs of associated activities;
   (ix) Natural bridges;
   (x) Surface depressions;
   (xi) Springs;
   (xii) Rocks, soil types or minerals characterized by mechanical weakness and/or a tendency towards dissolution or collapse, such as limestone, dolomite, gypsum, anhydrite, halite, terra rossa soils, lavas, weakly cemented clastic rocks, coal, or ores;
   (xiii) Non-conformities in soluble rocks;
   (xiv) Altered bedrock.
(b) Features causing additional bedrock instability:
   (i) Swelling rocks and shales;
   (ii) Potential displacement planes determined by unstable or mechanically weak subsurface layers;
   (iii) Faults and fracture zones and associated complex fracture systems.

2.10. The detection of most types of undesirable subsurface conditions is expected to result from the standard site characterization activities (see paras 2.11–2.23). However, it might be difficult to specify the criteria for exploration, testing and analysis for some undesirable conditions to ensure that the investigation programmes cover all abnormal subsurface conditions. For this reason, the recommendations in Section 3 of this Safety Guide should be followed to address any undesirable subsurface conditions, for which the potential of occurrence has been indicated in standard site characterization. Investigation programmes for complex subsurface conditions should include prediction, detection, evaluation and treatment.

Site characterization stage: Verification

2.11. In the verification stage, it is assumed that the generalized layout and foundation loads are established and the primary geotechnical and geological characteristics of the site are known (based on the investigations at the site selection stage). In addition to the features stated in para. 2.7(a), the following factors should, among others, be considered in the evaluation to take into account normal conditions, geotechnical hazards and other extreme conditions:

(a) Spatial information relating to the continuity, extent and geometrical arrangement of the subsurface materials and discontinuities (stratigraphy and geological structure), with reference to the site layout;
(b) Identification of other undesirable subsurface characteristics (see paras 2.9, 2.10 and 3.2–3.15), such as cavity zones, swelling rocks and shales, collapsing soils or soluble rocks, the occurrence of gas pockets, and potential displacement planes determined by unstable or mechanically weak subsurface layers;
(c) Liquefaction potential;
(d) Erosion potential;
(e) Feasible foundation types;
(f) Preliminary bearing capacity and other factors of foundation stability;
(g) Preliminary settlement ranges;
(h) Shoring needs for deep excavations;
(i) Dewatering needs;
(j) Excavation difficulty;
(k) Prior use of the site;
(l) Site preparation needs.

2.12. In the verification stage, the investigation programme should cover the site as a whole, but it should also be conducted on a smaller scale appropriate for the layout of the nuclear installation. The investigation programme should take into account site characteristics (e.g. compositional and structural heterogeneity within the subsurface materials) and their variability, available from the earlier stages of investigation, and the overall planned layout. The geotechnical site investigation phase should be carefully planned to ensure that it is structured, complete and sufficient to satisfy the expectations of the interested parties and to address any uncertainties. The following site investigation techniques and related points should be considered:

(a) Geophysical investigations, such as seismic refraction and/or reflection surveys. These investigations should be conducted to provide continuous lateral and depth information for the evaluation of subsurface conditions. Geological constraints should be considered in the interpretation of the survey results. The results should provide stratigraphical and structural geological information, information on the location of the groundwater table, and an estimate of wave velocities at the site. The geophysical investigations should be designed to optimally reflect the site characteristics (or the regional characteristics, if necessary) and their spatial variability. Drilling, coring and sounding should be used to complement the subsurface geophysical data (e.g. stratigraphical information) as well as to constrain and validate the interpretations of the geophysical data sets.

(b) Rotary borehole drilling, coring or sounding. These techniques are used to define the overall site conditions and to collect basic information about the subsurface materials. The method selected should be justifiable by the site conditions. Borehole drilling and coring involve extraction of cores or other samples for rock or soil qualification and laboratory testing. Sounding measures the resistance offered by the soil and is used in determining the soil profile. The recovered information typically includes rock and/or soil units and their stratigraphical order, the attitude and shape of the boundaries between the subsurface units (e.g. bedding, contact), the depth of the bedrock or load bearing stratum, and the presence and attitude of the structural elements (e.g. bedding, foliation, fractures, faults) within the subsurface materials. The investigations should be conducted along at least two intersecting survey lines that are oriented to capture the expected variation within the subsurface and have a common investigation hole at the line intersection. These investigations should be used to determine and map the soil profiles. The number of boreholes and their depths should be sufficient to verify that the site is suitable, with no unacceptable subsurface conditions. In addition to the extraction of cores or other samples for rock or soil examination and laboratory testing, the investigation holes can be used for the installation of instruments for long term in situ testing, stress monitoring and monitoring of the groundwater regime, taking into consideration the potential long term effects of these investigation holes on the site conditions. The possible effects of boreholes on the potable water regime should also be investigated (see SSG-92 [6]). If necessary, test pits or test tunnels should be used to facilitate a direct examination of the subsurface conditions.

(c)   In situ testing. In accordance with the subsurface conditions, in situ tests should be performed to measure the mechanical properties of the foundation materials. These tests should include in situ loading tests and piezometric measurements of the groundwater.

(d)   Laboratory testing. Laboratory testing consisting of index and classification tests sufficient to characterize the geomechanical properties of the strata and subgrade media should be conducted on rocks or soils. For cohesive and granular soil samples obtained during the drilling and/or coring operation, appropriate consolidation and shear strength testing should be conducted on the undisturbed samples (see para. 2.34) to allow an estimation of soil strength, stiffness, stress–strain responses and consolidation properties. Dynamic tests should be conducted in the laboratory to obtain the shear strain dependence of the shear modulus and the damping ratio of the soil.

2.13. The results from the site verification stage should provide the information necessary to establish broad design parameters and conclusions relating to the site and its characteristics. Therefore, the preliminary characteristics of the nuclear installation, such as the loads, the physical dimensions of the buildings, the preliminary structural engineering criteria and the preferred plant layouts, should be known at the end of the verification stage.

Site characterization stage: Confirmation

2.14. The purpose of the site confirmation stage is to confirm the results obtained in the previous stages and to ensure that the spatial and thematic coverage of the site characterization data and interpretations is sufficient for the purposes of final layout planning. The results of the site confirmation stage should address geotechnical parameter variability and uncertainty and should provide sufficient geotechnical data and parameters for the detailed design of the nuclear installation and the conduct of its safety assessment.

2.15. The content of the site characterization, in situ testing and laboratory testing programmes conducted in the confirmation stage should be planned on the basis of both the preliminary characteristics of the nuclear installation and the geotechnical characteristics of the site as identified in the previous stages. The plan should reflect the information necessary for the detailed design of the installation. Data validation and other necessary validations or verifications need to be undertaken in a timely manner to enable additional or repeat testing if deemed necessary. The results of these investigations should be used in evaluating the suitability of the preliminary layout and modifying it, as necessary. If planned layouts are changed and new locations are chosen, additional testing and investigations should be performed if necessary. The final confirmations should be consistent with the known geotechnical characteristics of the site and the final layout of the buildings on the site, including the final safety classification of the buildings (see para. 2.23).

2.16. In addition to refining the investigations conducted in the earlier stages (see paras 2.5, 2.7 and 2.11), the investigations should include sufficient in situ and laboratory tests to address the following:

(a) Detailed scrutiny of the potential for undesirable subsurface conditions, such as cavities, fracture systems and faults (see paras 3.2–3.15);
(b) A revised estimation of the bearing capacity of the soil and bedrock underlying the nuclear installation;
(c) A determination of the settlement of structures and the site amplification;
(d) Establishment of soil parameters and soil–structure interaction parameters (dynamic and static);
(e) Engineering assessments of liquefaction triggering and consequences;
(f) Evaluation of the preliminary site specific response spectrum (if needed).

2.17. A subsurface investigation and laboratory testing programme extending the programme described in para. 2.12 should be conducted at the site using a drilling scheme suited to the planned layout of the nuclear installation in order to adequately characterize the geotechnical conditions of the site. At sites of relatively uniform soil and bedrock conditions, a uniform grid method can be applied. In other cases, the grid spacing and orientation should be defined according to the extent, heterogeneity and geological structure of the subsurface units and discontinuities. Where heterogeneity and discontinuities are present, the usual investigation process should be supplemented with investigation holes at adequate spacings, depths and angles to permit detection of the site's geological and geotechnical features and their proper evaluation.

2.18. It may also be necessary to include complementary drilling in the investigation programme to either establish the soil model for studies of dynamic soil–rock structure interactions or to further delimit any undesirable subsurface conditions (see paras 2.9 and 2.10).

2.19. The necessary drilling depths depend on site conditions: drilling should be deep enough to allow the site conditions that would affect the structures, systems and components of the nuclear installation to be fully ascertained and to confirm the soil and rock conditions determined in previous investigations.

2.20. For sites characterized by very thick soils, drilling should be conducted to enable the evaluation of potential deep instability at the site and of potential effects associated with sloping sites.

2.21. If competent rock is exposed on the surface or encountered at a shallow depth, drilling should, at a minimum, penetrate to the greatest depth at which discontinuities or zones of weakness or alteration could affect the stability of the foundation. If such a depth cannot be unequivocally determined (e.g. owing to the continuity of steeply dipping weakness zones at large depths), drilling should be conducted to enable the discontinuities or zones of weakness or alteration to be adequately characterized so that technically justifiable evaluations of their significance for the nuclear installation can be made.

2.22. For sites of weathered shale or soft rock, drilling may need to penetrate deeper than is needed for the normal purposes of geotechnical design in order to facilitate site amplification, collapse and subsidence evaluations.

2.23. The distinction between items important to safety and other items should be considered when defining the details of the site investigations. The subsurface investigation and testing programme for structures not important to safety should follow relevant local, national or international codes and standards for conventional planning and building and proven engineering practices. Depending on the site characteristics, drillings may be necessary at the planned locations of buildings not important to safety. At least one investigation hole should be drilled at the planned location of every structure important to safety.[2] Where conditions are found to be variable, the number and spacing of drillings should be sufficient to obtain a clear definition of the changes in the soil and rock properties.

Pre-operational stage

2.24. Geotechnical investigations, studies and monitoring should be continued after the start of construction of the nuclear installation and until the start of operation in order to complete and refine the assessment of the site characteristics by incorporating geological and geotechnical data newly obtained during the excavation and construction of the foundations. As subsurface material is

---

[2] Some States define a minimum of three investigation holes for every structure important to safety [7].

exposed during and after foundation excavations, it should be carefully observed and mapped for comparison with the assumed design conditions and confirmed with the design itself. For offshore or inland sites with complex groundwater conditions, supplementary investigations of the groundwater regime may be necessary. Deformation features (e.g. faults; potential soft zones or soft interbeds in rocks; folds or joints; lateral compositional changes; materials susceptible to volume change; other features of engineering significance) discovered during construction should be carefully assessed to ensure that the safety objectives are not compromised.[3] If necessary, in situ tests may also be performed in the base of the excavation. The existing ground model should be validated and verified or it should be revised to reflect any new information.

2.25. The data obtained on actual performance in settlements and deformations due to structural loads should be used to verify the predicted behaviour of the foundations. Since the construction sequence is generally long, these monitoring data should be used to revise the settlement models and the soil properties on the basis of actual performance, if needed.

Operational stage

2.26. Selected geotechnical investigations and monitoring of geotechnical parameters are performed over the lifetime of the installation to confirm the conditions; to demonstrate the continued validity of the design basis, safety assessment and periodic reviews; and to support future reassessment, if necessary. During the operation of a nuclear installation, the settlement of structures and the displacement or deformation of foundations and items important to safety, as well as parameters such as the level of the water table and its seasonal fluctuations, should be monitored and compared with predictions to enable an updated safety assessment to be made. The parameters to be measured, the type of records to be obtained, the measurement intervals and all site evaluation activities to be conducted in the operational stage should be described in a maintenance and monitoring programme and assessed as part of the periodic safety review. Recommendations for the operational stage are provided in Section 5.

---

[3] Additional information about the significance of such findings can be found in Ref. [8].

SOURCES OF GEOTECHNICAL DATA FOR THE SITING OF NUCLEAR INSTALLATIONS

2.27. Data collected during geotechnical investigations allow informed decisions to be made concerning the nature and suitability of the subsurface materials. The sources of data are as follows:

(a) Historical and current documents and data sets;
(b) In situ investigations and tests;
(c) Laboratory tests.

Historical and current documents and data sets

2.28. The geotechnical investigations necessitate an understanding of the general geology of the area of interest. This should be obtained by means of field reconnaissance and a review of available historical and current documents. The site review should include references to internationally acknowledged scientific literature within the corresponding discipline and ensure an adequate interpretation and evaluation of the available data. The appropriate documents may include the following:

(a) Geological reports and other relevant literature;
(b) Geotechnical reports and other relevant literature;
(c) Satellite imagery and aerial photographs;
(d) Digital elevation models (e.g. light detection and ranging (LIDAR) method);
(e) Three dimensional models of the subsurface;
(f) Topographical maps;
(g) Geological maps and cross-sections, including soil and bedrock;
(h) Engineering geological maps and cross-sections;
(i) Geophysical maps and cross-sections;
(j) Hydrogeological maps, hydrological and tidal data, flood records, and climate and rainfall records;
(k) Water well reports and water supply reports;
(l) Oil and gas well records;
(m) Mining history, old mine plans and subsidence records;
(n) Indications for mineral resources and records of exploration history;
(o) Seismic observational (instrumental) data and historical earthquake and paleoseismic records, and relevant seismological studies;
(p) Contemporary accounts of landslides, floods, earthquakes, subsidence, slow bedrock movements and other geological events of significance;
(q) Records of the performance of structures and facilities in the vicinity.

2.29. Other possible sources of information should also be considered, such as observations, reports, publications, theses and models available from individual observers, geology and engineering departments of colleges and universities, government geological surveys and engineering authorities, work done by other persons in the vicinity of the site, and observations made at quarries in operation.

In situ investigations and tests

2.30. Geophysical tests, geotechnical tests and hydrogeological tests are available for soils and rocks. While these three types of test should be performed, their extent can vary according to the scale and goal of the investigation and the information already available (see paras 2.28 and 2.29).

2.31. Geophysical tests provide estimates of the continuation and consistency of the stratigraphy. In the domain of elastic deformation, these tests also allow data or information to be derived by back analysis of the test results. Geophysical tests generally have a large spatial coverage (in terms of depth and surface area) and provide rough estimates of parameters sufficient for the purposes of site evaluation (e.g. the thickness of the layers and the parameters defining their mechanical properties). The tests should include some of the techniques shown in Table 1. The tests should be selected in accordance with best practices, taking into account the subsurface conditions. Geophysical tests can be verified or complemented by the subsequent in situ tests. Complementary data sets may be combined to provide a robust characterization and understanding of ground conditions.

2.32. Geotechnical tests address the near surface area (to a depth of at least two times the shorter dimension of a structure's base, or to a depth where the change in the vertical stress due to applied loads during or after construction is less than 10% of the effective in situ overburden stress). If competent rock is encountered at lesser depths, borings should penetrate to the greatest depth where discontinuities or zones of weakness or alteration can affect foundations and they should penetrate at least 6 m into sound rock.[4] The tests can be performed using many different techniques, such as by means of boreholes or working directly from ground level. A list of some techniques for geotechnical investigations of soil and rock samples is provided in Table 2. The appropriate tests should be selected and conducted, taking into account the subsurface conditions. In some cases (e.g. when developing seismic site response characteristics), geotechnical testing of samples taken deeper in the soil profile is needed.

---

[4] More details can be found in Ref. [7].

2.33. Hydrogeological tests identify the characteristics, behaviour and distribution of the groundwater, its direction of flow, and its interaction with surrounding geological formations. Hydrogeological tests determine the filtration parameters of subgrade soils.

Laboratory tests

2.34. Laboratory testing should be conducted on the samples obtained using in situ investigation methods. The recovery of good undisturbed samples is important to the overall success of the laboratory testing. The treatment of samples after collection is as significant to their quality as the procedure used to obtain them; therefore, sampling should be performed in accordance with established procedures and practices. Handling, field storage and transport to the laboratory should be given careful attention. Sampling should be performed by means of pits, trenches or excavations and by in-hole methods. It may be necessary in certain circumstances to freeze (or otherwise preserve) 'cohesionless' soils to obtain undisturbed samples, and the effects of such preservation techniques on the results should be considered.

2.35. The purpose of laboratory testing is to supplement and confirm the in situ test data in order to characterize the soil and rock at the site fully and correctly over the whole range of expected strains. The material damping ratio of the soil, for example, as well as other properties for large strains, are not easily obtainable by in situ tests. All phases of the site investigation and the associated field and laboratory testing should be carefully planned and implemented so that the properties of soil and rock can be realistically assessed with an uncertainty level compatible with the accuracy requested for design.

2.36. The testing programme should identify and classify soil and rock samples that adequately represent the geological and geotechnical composition and properties within, and their variation across, the site. Their physical properties and engineering characteristics should be obtained from published data or by measurement. The laboratory tests should be conducted in conditions adequately representing the conditions of the site. A list of techniques for laboratory investigations of soil and rock samples and their purposes is provided in Table 3.

2.37. Site characterization parameters for use in the design profile should be carefully derived from the results of in situ tests (see paras 2.30–2.33) and laboratory tests. Any discrepancies between the results of in situ tests and laboratory tests should be investigated and reconciled.

16

Reporting

2.38. The results of the geotechnical investigations and the consequent site characterization should be documented in a detailed geotechnical report in accordance with the investigation and monitoring plans. This report should be compiled at the end of the confirmation stage and updated during the pre-operational and operational stages. In some circumstances, such as a large ground investigation, it may be beneficial to have separate reports with constrained scopes. The reports should include the following items:

(a) A description of the investigation programme and its basis;
(b) The layout of the planned buildings;
(c) Descriptions of the site geomorphology, including digital elevation models or other topographical data;
(d) The results and interpretations of geophysical surveys, including maps and cross-sections;
(e) Spatial information about the conducted drillings, including drilling based cross-sections;
(f) Geological maps and profiles;
(g) Engineering geological classifications, maps and profiles;
(h) Drilling logs and test pit logs;
(i) The results of in situ testing;
(j) The results of laboratory testing;
(k) Descriptions and results of laboratory analyses;
(l) Descriptions of the groundwater regime and the physicochemical, physical and chemical properties of the groundwater;
(m) Descriptions of potentially undesirable subsurface characteristics and/or unstable conditions;
(n) Documentation of the magnitudes and sources of uncertainties related to each stage of data collection.

TABLE 1. EXAMPLES OF TECHNIQUES FOR GEOPHYSICAL
INVESTIGATIONS OF SOIL AND ROCK MEDIA

Type of test	Parameter	Area of application	Remarks
Seismic refraction and reflection	Deformation propagation time	Site categorization	— For surface investigations and vertical sections — Most suitable if the velocity increases with depth and the rock surface is regular
Cross-hole seismic test	Dynamic elastic properties (shear wave and compression wave velocities)	— Site categorization, obtaining of velocities for particular strata, dynamic properties, rock mass quality — Results are used for seismic site response and soil–structure interaction analyses, liquefaction triggering assessment, and design of foundations	— For deep investigations — One hole is needed for emission and one hole is needed for reception of the seismic waves
Uphole and downhole seismic test	Dynamic elastic properties (shear wave and compression wave velocities)	— Site categorization, obtaining of velocities for particular strata, dynamic properties, rock mass quality — Results are used for seismic site response and soil–structure interaction analyses, liquefaction triggering assessment, and design of foundations	— For deep investigations — Measurements only need a single hole

TABLE 1. EXAMPLES OF TECHNIQUES FOR GEOPHYSICAL
INVESTIGATIONS OF SOIL AND ROCK MEDIA (cont.)

Type of test	Parameter	Area of application	Remarks
Nakamura method	Low level (ambient noise) vibrations	— Site categorization, obtaining of velocities for particular strata, dynamic properties, rock mass quality — Results are used for seismic site response and soil–structure interaction analyses, liquefaction triggering assessment, and design of foundations	— Horizontal to vertical spectral ratio is calculated — Passive seismic method to determine the resonant characteristics of a site (boring is not needed)
Electrical resistivity	— Electrical resistance or conductivity — Liquid table content	Internal erosion, location of saltwater boundaries, clean granular and clay strata, rock depth, and underground mines (detected via measured anomalies)	For deep or surface investigations
Nuclear logging	Water content, density	Settlements, liquefaction, foundations	Necessitates expensive logging techniques
Microgravimetry	— Residual anomaly — Acceleration due to gravity	Sinkholes; heterogeneities including faults, domes, intrusions, cavities and buried valleys (detected via measured anomalies)	
Ground penetrating radar	Reflections of electromagnetic radiation	Cavities, deformation zones, open and water-filled fractures	
Magnetic techniques	Magnetic field intensity	Site categorization, areas of humidity	Identification of surface lineaments, maintenance of dykes and dams

TABLE 1. EXAMPLES OF TECHNIQUES FOR GEOPHYSICAL
INVESTIGATIONS OF SOIL AND ROCK MEDIA (cont.)

Type of test	Parameter	Area of application	Remarks
Spectral analysis of surface waves	Dispersive character of seismic surface waves	Site characterization, subsurface composition and structure	Used to determine the variation in shear wave velocities with depth within layered systems
Microtremor array measurement	Dispersive character of seismic surface waves	Site characterization, subsurface composition and structure	Similar to seismic analysis of surface waves but uses passive sources and seismic noise
Multichannel analysis of surface waves	— Surface wave geophysical method — Shear wave velocity variations below the surveyed area	Site characterization, subsurface composition and structure	Uses various types of seismic source

TABLE 2. EXAMPLES OF TECHNIQUES FOR IN SITU GEOTECHNICAL
INVESTIGATIONS OF SOIL AND ROCK MEDIA

Type of test	Type of material	Parameter	Area of application	Remarks
Flat jack test	Rock	In situ normal stress	Deformability, convergence	Questionable results in rock with strongly time dependent properties
Hydraulic fracturing test	Rock	In situ stress state	Deformability, convergence	Affected by anisotropy of tensile strength

TABLE 2. EXAMPLES OF TECHNIQUES FOR IN SITU GEOTECHNICAL INVESTIGATIONS OF SOIL AND ROCK MEDIA (cont.)

Type of test	Type of material	Parameter	Area of application	Remarks
Direct shear stress test	Rock	Shear strength	Stability problems, foundations	Usually needs a sufficient number of tests for statistical control
Plate bearing test	Clay, sand, gravel, rock	Reaction modulus	Compaction control, settlement, foundations	For excavations and embankments
Pressure meter test	Clay, sand, gravel, rock	Elastic modulus, compressibility	Settlement, bearing capacity	Needs a preliminary hole
Hydro tests (pumping test, injection test, slug test, pulse test)	Clay, sand, gravel, soil, fractured rock	Field permeability	Transmissivity of soil, settlement	Needs a preliminary hole and piezometers
Vane shear test	Soft clay	Shear strength	Bearing capacity, slope stability	Not suitable for silt, sand or soils with appreciable amounts of gravel or shells
Static cone penetration test	Clay, sand, gravel	Cone resistance, undrained cohesion, shear strength	Settlement, bearing capacity	Includes cone penetration test
Cone penetration test	For all but very strong soils	Side friction and point resistance, shear wave velocity, pore water pressure, relative density	Provision of detailed information on stratigraphy; shear strength; liquefaction; site response; soil–structure interaction; foundations	— No samples recovered — Applicable in fine and coarse soils with an average diameter of grain less than 20 mm

TABLE 2. EXAMPLES OF TECHNIQUES FOR IN SITU GEOTECHNICAL INVESTIGATIONS OF SOIL AND ROCK MEDIA (cont.)

Type of test	Type of material	Parameter	Area of application	Remarks
Seismic cone penetration test	For all but very strong soils	Measurement of small strain velocities	Provision of detailed information on stratigraphy; soil velocity; site response; soil–structure interaction; foundations	No samples recovered
Active gamma cone penetration test	Clean sands	Density	In situ soil density	No samples recovered
Standard penetration test	Soils and soft rock	Blow counts	Provision of detailed information on stratigraphy; site response; soil–structure interaction; foundations; settlement	— Applicable in fine and coarse soils with an average diameter of grain less than 20 mm — Not suitable for boulders or hard rocks
Gamma–gamma borehole probe	Rock and soil	Density	Continuous measure of density	

TABLE 2. EXAMPLES OF TECHNIQUES FOR IN SITU GEOTECHNICAL INVESTIGATIONS OF SOIL AND ROCK MEDIA (cont.)

Type of test	Type of material	Parameter	Area of application	Remarks
Rock coring	Rock	— Lithology, discontinuity density, orientation and properties — Measurement of rock quality designation used for various empirical correlations	Detailed information on stratigraphy; rock structure and integrity; slope stability; foundations	Can be further used for laboratory tests, lithological and structural characterization, and rock mass classification (Q value)
Overcoring test	Rock	In situ stress state	Deformability, convergence	Difficult to implement in highly fractured rock
Dilatometer or Goodman Jack	Rock and soil	Young's modulus (E) in lateral direction	Settlement, foundations	
Dynamic cone penetration test	Clay, sand, gravel	Cone resistance, relative density	Liquefaction, settlement, foundations	Includes standard penetration test
Large penetration test; Becker penetration test	Gravelly soil	Cone resistance, relative density	Liquefaction, settlement, foundations	

TABLE 3. EXAMPLES OF TECHNIQUES FOR LABORATORY INVESTIGATIONS OF SOIL AND ROCK SAMPLES

Type of test	Type of material	Parameter	Characteristics investigated	Purpose
Fall cone test, Casagrande test	Clayed soil	Water content (through liquidity and plasticity indices)	Soil index and classification	Atterberg limits, compressibility, plasticity
Sieve, hydrometer	Coarse grained soil mixtures	Grain size characteristics, percentage of fines and their consistency limits, mean grain size, uniformity coefficient, minimum and maximum void ratio, particle angularity, sphericity and specific gravity	Index properties	Liquefaction, settlement, foundations
Dietrich–Frühling gasometer	All soils	Carbonates content	Physical and chemical properties of soils	Soil classification
Physical and chemical analysis of soil	All soils	Salt content	Physical and chemical properties of soils	Influence on permeability
Petrological (thin section) study of rocks	Rock	Identification of undesirable constituents in rock	Identification of minerals, their texture and other special features	Identification of compositional and microstructural variation for suitable treatment of foundation

TABLE 3. EXAMPLES OF TECHNIQUES FOR LABORATORY INVESTIGATIONS OF SOIL AND ROCK SAMPLES (cont.)

Type of test	Type of material	Parameter	Characteristics investigated	Purpose
Proctor test, gammametry, American Society of Testing and Materials test (relative density)	All soils	Humid and dry densities, water content, saturation ratio, relative density	Consolidation, bearing capacity	Settlement, consolidation, bearing capacity
Oedometer	All soils	Oedometric, Young's modulus, consolidation coefficient	Consolidation, permeability characteristics	Settlement, consolidation
Shear test box, triaxial compression test	All soils	Young's modulus, Poisson's ratio, cohesion and friction angle, underdrained and undrained conditions	Shear strength, deformation capability of soil	Bearing capacity
Chevron bend, Brazilian test	Rock	Mode I fracture toughness	Mechanical properties	Rock mechanical characterization
Punch-through-shear test	Rock	Mode II fracture toughness	Mechanical properties	Rock mechanical characterization
Cyclic simple shear, torsional shear, dynamic triaxial test	All soils	Undrained cyclic shear strength, dynamic Young's modulus, Poisson's ratio, internal damping, pore pressure, G–γ and η–γ curves	Dynamic characteristics of soils	Liquefaction, settlement, site response, soil–structure interaction, foundations

TABLE 3. EXAMPLES OF TECHNIQUES FOR LABORATORY
INVESTIGATIONS OF SOIL AND ROCK SAMPLES (cont.)

Type of test	Type of material	Parameter	Characteristics investigated	Purpose
Uniaxial and/or triaxial compression test	Rock	Young's modulus, Poisson's ratio, unconfined compression strength and cohesion friction parameters of intact rock	Mechanical properties	Rock mechanical characterization
Point load test	Rock	Unconfined compression strength of intact rock	Mechanical properties	Rock mechanical characterization
Direct and/or indirect tensile strength test	Rock	Tensile strength of intact rock	Mechanical properties	Rock mechanical characterization

GEOTECHNICAL CONSIDERATIONS FOR THE SITING OF NUCLEAR INSTALLATIONS

Parameters of the geotechnical profiles

2.39. The programme of in situ investigations and laboratory testing that is implemented to obtain information on the relevant subsurface material properties and to aid in the definition of the subsurface model should result in a distribution of values of the geotechnical parameters. At this point, on the basis of the available information, a set of representative parameters that are most suitable for use in the models for geotechnical analyses should be selected. In these analyses, the effects of uncertainties in the geotechnical parameters on the variability of the analytical results should be determined by means of parametric studies. In these parametric studies, the state dependency (e.g. density, stress, strain, stiffness) of the responses should be considered.

2.40. The selected set of parameters should be determined in order to perform the geotechnical evaluation necessary for the construction of a nuclear installation. The profile may be defined as a geometrical and mechanical description of the subsurface materials in which the best estimates and ranges of variation for the characteristics of the foundation materials are determined and described in a way that is directly applicable to the subsequent analysis. The profile should include the following:

(a) The geometrical description (e.g. subsurface stratigraphical descriptions, lateral and vertical extents, number of layers, layer thicknesses, layer slopes);

(b) The physical and chemical properties of soil and rock and the parameters used for classification;

(c) Primary (or pressure) wave (P-wave) and secondary (or shear) wave (S-wave) velocities (V_p and V_s, respectively), stress–strain relationships, static and dynamic strength properties, strain dependent modulus degradation and damping relationships, consolidation, permeability, and other mechanical properties obtained by in situ tests and/or laboratory tests;

(d) Characteristics of the groundwater table, the design level of the water tables and the maximum water level in the case of design basis flooding and other conditions (e.g. runoff inundation or erosion, depth to groundwater, spring or groundwater discharge within or near the site).

2.41. Even though conceptually the profile is unique to a particular site, various related design profiles for different uses or assessments should be adopted to allow for different hypotheses in the analysis. These include design profiles for the assessment of the following:

(a) Site specific response spectra;
(b) Liquefaction engineering;
(c) Stresses in the foundation ground;
(d) Foundation stability;
(e) Soil–structure interaction;
(f) Settlements and heaves;
(g) Stability in earth structures;
(h) Earth pressure and deformations or displacements in buried structures.

Seismic site categorization

2.42. For the purpose of seismic site response analyses, the following categorization of subsurface media can be used:

— Type 1 sites: $V_{s,30\,m} > 1100$ m/s;
— Type 2 sites: 1100 m/s $> V_{s,30\,m} > 360$ m/s;
— Type 3 sites: $V_{s,30\,m} < 360$ m/s.[5]

This site categorization is based on the assumption that the shear wave velocity (V_s) smoothly increases with depth. If this assumption is not fulfilled (i.e. V_s decreases or abruptly increases with depth in the upper 30 m, or if there is a strong impedance contrast at any depth), specific analyses including site response analyses should be performed in accordance with best practices, regardless of the site type. If this site categorization is not applicable, soil investigations should be performed to determine the soil type for the site or to provide comprehensive data for further analyses.

2.43. Regardless of the site type, if the value of $V_{s,30\,m}$ adopted as part of the probabilistic seismic hazard assessments is not in conformance with the shear wave velocity profile of the site, then site response analyses (incorporating a suitable deconvolution scheme as applicable to the approach used) should be performed.

Free field seismic response and site specific response spectra

2.44. The seismic input level that should be considered is the SL-2 level of seismic vibratory ground motion hazard, as specified in SSG-67 [3].

2.45. Seismic site response analyses under free field conditions should be performed for Type 2 and Type 3 sites (see para. 2.42) or when the site specific conditions differ from the ground motion model reference conditions. Site response analyses provide input parameters for the assessment of cyclically induced displacements

---

[5] $V_{s,30\,m}$ can be estimated using the following equation (although other acceptable estimations may be used) from the representative small strain ($<10^{-4}$%) shear wave velocity profile of the site in its natural conditions before the execution of site works:

$$V_{s,30\,m} = \frac{30\ \text{m}}{\sum_{i=1}^{n} \Delta t_i}$$

where Δt_i is the travel time of the shear wave in the ith layer ($\Delta t_i = \Delta H_i / V_{s,i}$) and ΔH_i is the thickness of the ith layer in the upper 30 m.

and deformations (including those for soil liquefaction engineering) as well as for soil–structure interaction analyses. Additionally, the site response analyses should provide site specific response spectra. At a minimum, data on the following should be collected:

(a) The input ground motion (derived by means of the procedures described in SSG-9 (Rev. 1) [2]).

(b) An appropriate model of the site, based on:
  (i)   The geometrical description of the soil and rock layers;
  (ii) The velocities of the P-waves and S-waves in each layer;
  (iii) The relative density and the density in each layer;
  (iv) Strain dependent modulus degradation and damping relationships, which describe the apparent reduction in shear modulus G, and the corollary increase in internal damping ratio ξ of the soil layers with increasing shear strain γ levels (i.e. G–γ and ξ–γ curves).

(c) For deep soil deposits in which wave velocities increase smoothly with depth, the change in the parameters described in subparagraph (b) with increasing confining stress and/or depth.

2.46. Depending on engineering practices, the seismic scenario compatible outcrop motions recorded at a reference site (e.g. a site with a reference $V_{s,30\,m}$ value) should be selected from available ground motion databases (i.e. databases that include strong motion recordings and associated metadata). These outcrop input motions should be chosen in accordance with the event type, the event magnitude, the distance to the seismic source, the directivity effects, and the characteristics and elevation of bedrock in the soil profile, all of which govern the amplitude, frequency content, duration and other relevant strong motion characteristics. If necessary, these records should be scaled in amplitude or duration or modified in spectrum to match the target seismic scenario, while maintaining consistency with the strong motion characteristics. Synthetic records can also be tailored using a combination of Fourier amplitude spectra and random vibration theory.

2.47. In the case of an input ground motion provided as a free field outcrop motion, a deconvolution of the outcropping input motion to a point within the soil column should be performed (e.g. at the foundation level, at a point of interest for liquefaction assessment). Deconvolution may result in a reduction in the intensity of ground motion within the soil column compared with that of the outcrop. This reduction should be carefully reconsidered and justified by means of parametric studies.

2.48. Alternative methods to assess the idealized layered soil–rock systems include wave mechanics, finite element, finite difference, discrete element and hybrid methods. To assess the site response, models with the following properties are acceptable:

(a)   A viscoelastic soil system overlying a viscoelastic half space;
(b)   A horizontally layered system;
(c)   Materials that dissipate energy by internal damping;
(d)   Vertically propagating body waves (shear and compression waves).

2.49. The equivalent linear models of soil constitutive relationships should be consistent with the strain level induced in the soil profile by the response to the input ground motion. If non-linear models are used, the strain dependent modulus degradation and damping responses should be captured as part of the constitutive model implemented.

2.50. Uncertainties in the mechanical and dynamic properties of the site materials should be considered through parametric studies. A single set of soil profile parameters should not be assumed to be conservative for all the scenarios considered (e.g. a conservative profile for deconvolution might not be conservative for the site response analysis).

2.51. When the site is in the near field of a seismic source, the site response model should be carefully determined so that the frequency content of the input motion affected by the earthquake mechanism may be appropriately assessed, taking into consideration the directivity effects. For these cases, time histories should be selected to include pulse-like motions in the ensemble of input motions.

2.52. In seismic response analyses of Type 3 sites, significant de-amplification in acceleration levels may be observed. In such cases, assessments supported by engineering judgement based on parametric studies should be considered.

3. GEOTECHNICAL HAZARDS IN SITE EVALUATION FOR NUCLEAR INSTALLATIONS

3.1. Requirement 22 of SSR-1 [1] states that **"Geotechnical hazards and geological hazards, including slope instability, collapse, subsidence or uplift,**

and soil liquefaction, and their effect on the safety of the nuclear installation, shall be evaluated."

UNDESIRABLE SUBSURFACE CONDITIONS AT NUCLEAR INSTALLATION SITES

Prediction of undesirable subsurface conditions

3.2. Potentially undesirable subsurface conditions should be investigated. An understanding of the regional and site geology can provide indications of potential ground collapse. This investigation should include consideration of soluble rocks (which are usually either sedimentary rocks, including carbonate types (mainly limestone and dolomites that are appreciably soluble in water or in weakly acidic solutions), or evaporites, of which halite, gypsum and anhydrite are the most common). The current size and future evolution of the size of potential cavities or underground solutions are governed by geological factors and environmental factors, both of which should be considered. The geological factors include the potential for buried channels, the stratigraphical sequence, the characteristics of the rock type and the properties of the rock mass. The environmental factors include surface water and groundwater hydrology as well as climate, including the effects of climate change.

3.3. The mechanical stability of the bedrock is governed by the stress state, the properties of the rock mass and the discontinuities transecting the rock mass at all depths of interest. As the discontinuities might define complex patterns and networks, their occurrence, orientation and properties should be investigated. Prediction of the future evolution of discontinuities should involve a review of the deformation history of the site and its wider surroundings, with specific focus on the presence of deformation zones (e.g. faults, shear zones) and their character. The review should consider the potential for slow movements between juxtaposed bedrock blocks due to glacial rebound, tectonism, groundwater extraction and other industrial activities. Capable faults are required to be identified and evaluated (see Requirement 15 of SSR-1 [1]).

Detection of undesirable subsurface conditions

3.4. The investigation programme at a site, as outlined in Section 3, should provide for the detection of subsurface cavities and allow for their extent and formation to be evaluated. The possibility of the detection of areas susceptible to ground collapse and the resulting complications should be considered in all aspects of the investigation programme. The conventional methods of site investigation are applicable, including geophysical surveys, remote sensing, aerial surveys, drilling, sampling, excavation, borehole logging and hydraulic pressure tests.

3.5. If the presence of subsurface cavities is suspected at a site, the initial subsurface investigation programme to locate cavities should aim to identify their size and spatial distribution. Some geophysical methods are useful for the detection of geophysical anomalies, which could correspond to potential subsurface cavities. Such methods include surface electrical resistivity profiling, microgravimetry, low resolution seismic refraction surveys, seismic fan shooting and ground penetrating radar. If detected, geophysical anomalies should be confirmed by drilling (and remote visual inspections if necessary) to determine their properties, for example the depth, size and geometry of cavities.

3.6. Geophysical methods that can be used as preferred resolution survey techniques in determining the depth, size and geometry of subsurface cavities include cross-hole seismic survey, cross-hole radar methods, electrical resistivity survey, acoustic resonance with a subsurface source, microgravimetry, high resolution seismic refraction, high resolution seismic reflection, surface wave method, ground penetrating radar methods, and suspension P-wave and S-wave logging. Several of these methods should be applied, in conjunction with tomographical techniques, for cross-validation.

3.7. Geophysical investigations should be carefully planned and, typically, implemented in conjunction with drilling and sampling techniques that enhance their effectiveness. The result of an investigation programme to detect potential subsurface cavities and, if present, to define subsurface cavities and their potential patterns should be a map or a cross-section showing the cavities and their relationships to the structures, systems and components on the site.

3.8. It might not be possible or practicable to detect and delineate every possible cavity or solution filled feature at the site. Consequently, a decision should be made regarding the largest possible undiscovered cavity that would be tolerable, based on the potential effects of such cavities on the performance of structures, systems and components important to safety.

3.9. The detection of significant mechanical discontinuities in the rock mass should be in accordance with the site investigation procedures (see para. 2.12).

3.10. Evaluation of the significance of bedrock discontinuities should involve characterization of the geometry, size, topological relationships and mechanical properties of the discontinuities. This characterization should enable an understanding of how these discontinuities are arranged into fault and fracture systems and networks. Such an understanding is necessary in evaluating the potential of these discontinuities to cause movements of bedrock blocks and faulting, including slip along the main slip surface of the fault, as well as secondary displacements in fractures spatially associated with the faults.

Evaluation and treatment of undesirable subsurface conditions

3.11. The greatest risk to the foundation safety of a nuclear installation, from a geotechnical perspective, is from the existence of filled or open cavities, solution filled features at shallow depths (relative to the size) and mechanical discontinuities below the foundation of the structures, systems and components at the site. The compressibility and the erosion potential of the natural filling material should be evaluated to determine their impact on bearing capacity, settlement and future erosion as a result of possible changes in the groundwater regime.

3.12. The stability of natural cavities and mechanical discontinuities below the foundation level should be considered. The size of the cavity and its depth, the patterns and properties of the associated mechanical discontinuities, the type of rock, and bedding inclinations above the cavity are primary factors that influence the stability of the cavity roof and the depth of foundation level under consideration. Changes in the vertical pressures due to structural loads or seismic events could cause instability of the cavity roof. In areas where the size and geometry of the cavity can be reliably determined, analytical methods such as finite element analysis and finite difference analysis should be used for the evaluation of the stability of the cavity. A site that is underlain by a potentially large and complex cavity system should be excluded, since the hazard posed by the cavity system is difficult to evaluate realistically.

3.13. For sites where complex subsurface conditions are encountered below the foundation level, the results of the stability evaluation should indicate the need for ground treatment to ensure the safety of the structures. Further recommendations on the improvement of foundation conditions at sites with complex subsurface conditions are provided in Section 4.

Improvement of surface conditions and subsurface conditions

3.14. If it has been found necessary to make improvements in the subsurface conditions owing to the risk of slope failure or other unfavourable soil or ground conditions, the improvements (e.g. jet grouting, ground cementation) should be designed and implemented during the ongoing stage of site characterization and/or site preparation and construction, and their effectiveness should be verified by in situ testing (see also paras 3.48, 3.49 and 4.17–4.20).

3.15. In areas subjected to slow differential movements of bedrock blocks (e.g. due to unevenly distributed glacial rebound), engineering countermeasures should be considered. In such cases, a layer of crushed rock can be used as a mitigation technique, and the movements should be monitored and assessed against well established and defined limits for maximum allowed movements.

NATURAL SLOPES ON OR NEAR SITES FOR NUCLEAR INSTALLATIONS

3.16. A natural slope is composed of rocks and/or soils. In rock slopes, the existence of weak parts, such as weak layers, lithological contacts and discontinuities (e.g. joints, faults), plays an important role in the stability of the slope. In soil slopes and weak parts in rock slopes, an increase of pore water pressure caused by heavy rainfall or earthquakes should be evaluated if the water table level is within the slope.

Slope stability

3.17. Slope stability assessment depends largely on the distance of the slope from the nuclear installation and site and on the potential outreach of the slope. Potentially hazardous slopes should be identified and evaluated in terms of such factors as distance from the site or installation, orientation, slope angle, height, geology and groundwater level, as well as any changes in these factors over time (e.g. additional units at the same site, settlements within the slope, glacial rebound, riverbank erosion, coastal erosion, groundwater changes and/or climate change). If a slope is determined to be distant enough that it would not affect any items important to the safety of a nuclear installation, the emergency planning zones or any other important site features, no further measures are necessary.

3.18. The stability of slopes in the vicinity of items important to the safety of a nuclear installation should be assessed with regard to the safety of the installation. In particular, the effects of earthquakes (e.g. ground motion, liquefaction, landslides, tsunamis) as well as the effects of heavy rainfall, flash floods and thawing permafrost should be considered in the assessment of slope stability.

3.19. For pseudo-static slope stability calculations, the methodology is based on the consideration of seismic effects as equivalent static inertial forces by means of seismic coefficients. To determine the equivalent static inertial forces, the seismic amplification in the slope should be based on a seismic loading distribution along the vertical direction of the slope. Peak ground acceleration can be used for the initial estimation of the inertial forces. However, a lower value might be acceptable, if justified by additional calculations and studies.

3.20. If the resulting safety factor is not greater than the specified minimum[6] (i.e. regulatory expectation), a dynamic response analysis based on the design seismic ground motion should be performed to evaluate the seismic effects more precisely. If necessary, the permanent displacements (i.e. residual deformation) should be evaluated to assess safety and stability in cases where the safety factor is close to unity. For sites on, or surrounded by, natural slopes, these evaluations are important for beyond design basis external events, and the results should be considered with respect to cliff edge effects for nuclear installations.

3.21. If natural slopes are credited as barriers against floods or tsunamis, the influence of ground erosion and related changes of material properties and slope geometry should be taken into account in the safety assessments and evaluations.

3.22. If a slope is deemed to be potentially unstable, a stability analysis should be performed. The stability analysis should consider factors such as slope angle, height, water content, groundwater level, reduced soil strength under seismic loadings, and other geotechnical conditions of the material of the slope, as well as the potential uncertainties associated with these factors due to the variability of the slope material (e.g. primary stratification of the sediments; see para. 2.40).

3.23. A conventional sliding surface analysis is usually performed to evaluate a safety factor against sliding failure. This method is based on a simple equilibrium of force and is valid for an external load such as gravity. However, for loads such as those generated by an earthquake, an additional evaluation should be conducted to determine the exact location of the expected sliding surface if it is different from the sliding surface determined using the minimum safety factor that considers only gravity and the residual shear strength of the slope. A three dimensional slope stability analysis might be needed to evaluate more realistically the stability of the slope and the impact of a failed slope portion.

---

[6] In slope stability calculations, the resulting safety factor calculated on the basis of the pseudo-static equilibrium should be at least 1.1. However, different national regulations and practices may specify a minimum safety factor as high as 1.5.

3.24. If the evaluation results in a safety factor low enough to indicate potential for a major sliding failure, suitable measures for stabilizing and strengthening the slope and/or for preventing any debris from reaching structures, systems and components important to safety should be designed and implemented. Alternatively, the layout of the nuclear installation site should be modified.

Measures for prevention and mitigation of slope failure

3.25. If a natural slope is assessed to be not sufficiently safe (by a safety factor and/or any other criteria, such as permanent displacements), measures for the prevention and mitigation of slope failure should be considered, such as the removal of the whole or part of the natural slope. If removal is deemed unreasonable, strengthening measures should be considered, such as lowering of the slope angle, soil nailing, rock bolting, grouting, anchors, piles and/or retaining walls.

3.26. Different mechanisms can be used to strengthen a slope with anchors (e.g. providing extra confining pressure to increase the strength of the slope material by pretension of the anchor, using the strength of the anchors to hold a sliding block after sliding is initiated). The mechanism selected should be supported by a quantitative comparison of the various options and should be agreed with the regulatory body.

3.27. Measures should also be considered to prevent any debris from reaching structures, systems or components important to safety. For example, a protective wall can be designed to stop the debris after an external event of a certain severity that might exceed the stability of the slope. The wall should be designed with consideration of the maximum and minimum size of the falling debris estimated to reach the wall. The design should ensure that the wall will withstand the loads of the debris and its impact as well as the earth pressure to be retained.

SOIL LIQUEFACTION ON SITES FOR NUCLEAR INSTALLATIONS

3.28. Soil liquefaction should be fully described using definitions of the soil behaviour and loading conditions (e.g. flow liquefaction versus cyclic softening, soil response to shear stresses, controlling stresses, onset of threshold strain levels, excess pore pressure ratio). This forms the basis of any liquefaction engineering assessments for a nuclear installation site. Such a basis should be established, and acceptable performance levels should be defined.

3.29. The most significant seismic design scenario adopted for liquefaction assessments might not necessarily be the same as that used for the assessment of overlying structural systems. A distant but larger magnitude seismic event with a lower intensity but longer duration (producing a larger number of equivalent stress cycles) may be more significant for liquefaction response.

3.30. Liquefaction engineering assessment procedures should be followed for beyond design basis external events, where the seismic input level is selected for a return period exceeding the SL-2 level of seismic vibratory ground motion hazard. The performance of items important to safety during and after beyond design basis external events should be evaluated against predefined acceptance criteria to avoid cliff edge effects.

3.31. The necessary data should be collected for the liquefaction engineering assessments. The following list presents relevant types of data:

(a) Historical performance data. Data available for soils with properties identical or similar to those at the site should be compiled and studied. Additionally, if available, the cyclic performance of the site during and after historical earthquake events should be documented.

(b) Soil profile. A detailed representative soil profile indicating the stratigraphical characteristics of each layer, with special emphasis on their spatial variabilities, should be developed.

(c) Groundwater regime. Piezometric and/or borehole water level data should be used to define the phreatic surface. The seasonal and situational fluctuations in the phreatic surface (e.g. fluctuations due to flooding, tsunami or climate change) should be conservatively considered in the assessments. Additionally, data from borehole pump and/or cone penetration tests with pore water pressure measurements can be used to determine the permeability parameters.

(d) Index properties. For coarse grained soil mixtures, tests including sieve and sedimentation, laser diffraction and/or hydrometer testing should be performed on soil samples to assess grain size characteristics. Samples should be collected to accurately represent the spatial variability of the site soil conditions. In addition to the percentage of fines and their consistency limits, mean grain size, uniformity coefficient, relative density and specific gravity are important properties useful for liquefaction engineering assessments.

(e) Standard penetration tests. There is significant variability in the equipment used and in the procedures and protocols adopted for standard penetration testing. To minimize this variability, such testing should be performed in conformance with standardized testing methodologies (e.g. those developed by the International Organization for Standardization and ASTM International). Additionally, to allow for possible test corrections, the equipment details (e.g. sampler type and dimensions, hammer type, cathead-rope-pulley system details (for non-automatic hammers), rod type, rod length, coupling type and dimensions, anvil-hammer, anvil-rod inclinations) should be fully documented. Either a calibrated standard penetration test hammer system should be used or direct stress wave energy measurements should be performed in situ, in conformance with standardized testing methodologies (e.g. those issued by the International Organization for Standardization and ASTM International). The field blow counts should be corrected to consider the variability in the procedures followed, the equipment used and stress states. Considering the spatial variability of standard penetration test blow counts, representative blow counts should be determined using deterministic or probabilistic methods. When gravelly soil layers are present, a large penetration test, a Becker penetration test or shear wave measurement results should be used for the assessments.

(f)   Cone penetration tests. The cone penetration test has an advantage over the standard penetration test in that it provides a continuous soil profile, allowing better judgement about the extent of liquefiable soil layers. However, unless customized systems are used, conventional cone penetration testing equipment does not allow soil sampling, so soil classification should be developed on the basis of sleeve friction and cone tip resistance data. Additionally, penetrability decreases with increasing soil density and grain size, which might limit the use of cone penetration tests in gravelly and/or cemented sandy soils. Under these circumstances, standard penetration tests and cone penetration tests should be performed either jointly or in combination with boring. For reliable assessments, calibrated cone penetration test equipment and sensors should be used.

(g)   Shear wave velocity measurements. Such measurements are a complementary tool for liquefaction triggering assessments. There are different shear wave velocity measurement techniques with different levels of accuracy. Downhole and cross-hole measurements include the drilling of boreholes and sampling. Non-invasive surface measurement techniques (e.g. seismic analysis of surface waves, multichannel analysis of surface waves) can also be considered but only provide a mean shear wave velocity value per layer of the soil profile. Seismic cone penetration test systems may also be considered to collect both shear wave velocity and cone penetration test data simultaneously. When possible, multiple independent field test data collection methods should be used to reduce the epistemic uncertainty in liquefaction triggering predictions.

(h)   Relative density. The in situ relative density of cohesionless soils should be evaluated on the basis of the standard penetration test blow counts and/or the cone tip resistances using justified correlations and the estimations should be compared with the test results of undisturbed samples. Conversely, clean soil samples (i.e. fines content < 5%) at the target relative density can be directly reconstituted in the laboratory, after estimating the minimum and maximum void ratios, for which standardized testing methods are available.

(i)   Undrained cyclic shear strength. The undrained cyclic shear strength of soils may be evaluated directly by means of cyclic loading tests performed in the laboratory on undisturbed (frozen) or reconstituted soil samples. Cyclic simple shear, torsional shear and triaxial tests, along with centrifuge models, are commonly employed in engineering practices to evaluate the undrained cyclic response of soils. The quality of the undisturbed samples or the method of sample preparation (e.g. reconstitution) for laboratory tests significantly affects liquefaction response and should therefore be considered in the interpretation of the assessment results. An alternative to laboratory based assessments is case history based semi-empirical methods for the evaluation of liquefaction resistance and post-liquefaction (residual) shear strengths.

(j)   Strain dependence of soil properties. For advanced dynamic analysis, strain dependent soil properties for each soil layer are needed to describe the apparent degradation in shear modulus and the increase in damping ratio with increasing shear strain levels, respectively (see para. 2.45).

(k)   Additional soil properties. Additional parameters (e.g. Poisson's ratio, critical state soil parameters) may be needed as part of more advanced assessments.

(l)   Seismic design parameters. At least one pair of moment magnitude and peak ground acceleration values are needed from deterministic seismic hazard assessment. Alternatively, peak ground accelerations deaggregated into moment magnitude bins, or peak ground acceleration levels corrected to a reference magnitude (and duration) are needed from probabilistic seismic hazard assessments.

(m)   Ground motion duration. The number of equivalent uniform stress cycles corresponding to the magnitude of the seismic design event should be determined. The magnitude of the seismic event is commonly used to assess the duration of seismic demand on the assumption that ground motion duration can be correlated, as a first approximation, to the number of cycles of the earthquake.

(n)   Cyclic stress ratio. The induced cyclic stress ratio at the depth of interest — which can be estimated by seismic site response analyses or by simplified procedures using site-response based soil mass participation factors — should be evaluated.

(o) Laboratory based cyclic resistance. For laboratory based assessments, the cyclic stress ratio versus the number of equivalent stress cycle curves that correspond to the triggering of liquefaction should be developed.

(p) Laboratory–field condition corrections. A set of correction factors to account for the differences between laboratory conditions and field conditions should be developed and justified.

(q) Additional seismic parameters. These parameters (e.g. design basis time histories) may be needed for more complex assessments.

3.32. Liquefaction engineering assessments should include, at a minimum, the following engineering evaluation steps:

(1) Liquefaction susceptibility and triggering;
(2) Post-liquefaction residual strength and overall post-liquefaction stability;
(3) Cyclically induced deformations and displacements;
(4) Consequences of induced deformations and displacements;
(5) Engineered mitigation (if necessary).

Liquefaction susceptibility

3.33. As part of susceptibility assessments, fully saturated clean sands, clean gravels (see para. 3.31(h)), non-plastic silts, and mixtures of these should be considered susceptible to liquefaction. Clean sands or gravels are defined as soils with a fines content of less than 5%. The lateral extent of the susceptible soil layers should be confirmed and studied in the overall stratigraphical context.

3.34. Mixtures of sands and/or gravels with plastic fines should also be assessed for susceptibility to liquefaction. For the susceptibility assessment of plastic fine grained soil mixtures, data on grain size, grain distribution, consistency limits and experimentally assessed pore pressure generation can be used. If soils are concluded to be susceptible to liquefaction, liquefaction triggering assessments should be performed.

Liquefaction triggering

3.35. Three approaches are used as part of liquefaction triggering assessments:

(a) Case history based semi-empirical approaches (see para. 3.41);
(b) Analytical approaches (see para. 3.42);

(c)   Advanced constitutive model based numerical approaches (see para. 3.43).

3.36. It is generally possible to compute a lower bound solution in all three approaches to liquefaction engineering assessments outlined in paras 3.41–3.43 by using conservative assumptions for the design profile parameters. For loose sands, a slight increase in the seismic stresses could bring the soil into an unstable condition, with possible large deformations, while in medium to dense sands, even a large increase in seismic stresses might only generate limited deformations, even if pore pressure buildup is 100%. Therefore, cliff edge effects should be considered.

3.37. For deterministic assessments, the safety factor against liquefaction triggering should be greater than the limit value considered for the calculation and should be consistent with the methods used (as prescribed by regulations or in accordance with standardized codes). For probabilistic assessments, a sufficiently low threshold frequency of liquefaction triggering should be established to satisfy performance targets.

3.38. Fulfilling the minimum safety factor or annual probability of liquefaction triggering might not guarantee acceptable displacement or deformation performance. Thus, the rest of the liquefaction engineering assessments (see paras 3.44–3.49) should be performed regardless of the liquefaction triggering evaluation outcome.

3.39. When liquefiable conditions exist within a soil layer, their volume should be estimated using resistance profiles measured in situ (e.g. beds, lenses, extended layers). The distribution of these liquefiable layers and their configurations, distances, hydraulic connections, and permeability contrasts, as well as the proximity of the drainage boundaries, should be considered in the liquefaction triggering assessment. If there are insufficient details in the data, the whole layer should be considered liquefiable.

3.40. Liquefaction triggering assessments should consider groundwater levels, which should be defined from piezometric measurements and should take into account groundwater fluctuations.

Case history based semi-empirical approaches

3.41. Semi-empirical approaches are based on deterministic or probabilistic assessment of liquefaction triggering case histories from historical events, where capacity versus demand terms are selected as an in situ test parameter (e.g. normalized standard penetration test blow counts, normalized cone penetration test blow counts, normalized Becker penetration test counts, normalized shear wave velocity) versus a normalized cyclic stress ratio. These semi-empirical approaches use the earthquake moment magnitude, fines content, non-linear shear mass participation factor and some basic soil parameters (e.g. unit weight, grain size, consistency limits).

Analytical approaches

3.42. The analytical approach used for liquefaction triggering assessments should comprise the following steps:

(1) Choose a set of representative accelerograms, consistent with the seismic design scenario, at the outcropping reference rock site.

(2) Deconvolute or convolute the outcropping reference surface motions to motions at the depth of interest and estimate induced cyclic shear stress histories through a set of seismic site response analyses.

(3) Convert the number of cycles of transient stress–time histories into equivalent uniform stress cycles.

(4) Develop the cyclic resistance ratio versus the number of equivalent uniform stress cycle curves through a set of cyclic laboratory tests.

(5) Assess the liquefaction triggering response by comparing the induced cyclic stresses with the cyclic resistance corresponding to the number of equivalent uniform stress cycles estimated earlier.

Advanced constitutive model based numerical approaches

3.43. A validated and calibrated constitutive model, capable of modelling the cyclic large strain response of fully saturated soils, should be incorporated into the non-linear, time step analysis to directly assess the buildup of pore pressure and the overall seismic response. As part of these assessments, effective stress based, time domain, coupled or decoupled analyses are usually performed to simulate strain and time dependent changes in soil stiffness and strength along with the buildup of pore pressure. The onset of liquefaction triggering can be directly identified under the cyclic loading defined by the set of input motions used. However, the results might vary considerably owing to the use of different input motions, different constitutive models and/or a different set of constitutive model parameters. Advanced dynamic analyses necessitate the calibration of many parameters that are difficult to identify in routine applications. The results should therefore be calibrated with case history based evaluations and should consider the uncertainties in the parameters used in the analysis.

Post-liquefaction residual strength and overall post-liquefaction stability

3.44. If it is concluded that soils could liquefy during the design basis seismic event, post-liquefaction residual strength and overall post-liquefaction stability assessments should be performed, taking into consideration the uncertainties associated with the parameters and methodology used. Semi-empirical, analytical and calibrated constitutive model based assessments can also be used to assess post-cyclic residual strength. Post-liquefaction stability assessments should include the applicable potential failure modes, including slope stability, bearing capacity, uplift, sliding and toppling, and others if relevant. These assessments should also consider earthquake aftershocks during transitional phases (e.g. before pore water pressures have fully dissipated) and all changes of soil states after the main shock (see para. 4.94), if applicable.

3.45. If post-liquefaction overall stability cannot be guaranteed, mitigation solutions should be engineered and implemented against soil liquefaction. In overall stability evaluations, an acceptable safety factor and/or acceptable deformation and displacement performance levels should be selected to comply with short term loading conditions.

Cyclically induced deformations and displacements and their consequences

3.46. When overall stability is achieved, cyclically induced deformations and displacements should be evaluated. Post-liquefaction differential settlements and their associated uncertainties should be assessed.

3.47. The consequences of cyclically induced deformations and displacements should also be assessed. The deformations and displacements should comply with acceptable performance criteria. Acceptable levels of performance with regard to preserving repairability, reducing overall damages, maintaining serviceability and/or minimizing out of service duration should be defined.

Engineering mitigation

3.48. If cyclically induced deformations and displacements do not fall within the acceptable performance levels described in para. 3.47, mitigation solutions should be engineered and implemented.

3.49. The engineering mitigation solutions for the unacceptable liquefaction hazard should be performed on the basis of applicability, effectiveness, ability to verify the reliability of the mitigation achieved, regulatory requirements and other concerns (e.g. costs, environmental issues).

4. GEOTECHNICAL CONSIDERATIONS FOR DESIGN AND SITE EVALUATION FOR NUCLEAR INSTALLATIONS

DYKES AND DAMS ON OR NEAR SITES FOR NUCLEAR INSTALLATIONS

4.1.   The term 'dyke' is used to describe a structure running along a watercourse, and the term 'dam' (or 'earth dam') applies to a structure used to create a water reservoir upstream or downstream from a nuclear installation.

4.2.   Before construction, in addition to classical geophysical and geotechnical tests, special attention should be paid to the soil and/or rock permeability of the site close to the areas of the foundations. Soil and/or rock permeability should be monitored throughout the operating lifetime of the installation.

4.3.   The design and evaluation of dykes and dams should consider all possible failure modes (including those dependent on pore pressure inside the embankment and on internal erosion caused by water seepage and flow inside the embankment).

4.4.   The design requirements for dykes and dams relating to the consequences of their failure that might impact the safety of the nuclear installation (e.g. due to the loss of cooling water) should be consistent with the design requirements for the installation itself, especially with regard to the evaluation of natural hazards (e.g. earthquakes, rainfall), including the return period for flooding.

4.5.   In addition to the usual methods of engineering design, a specific analysis should be performed to evaluate the relevant parameters of the structures (e.g. displacements, pore pressures), the values of which should be compared with those measured in situ at the different stages of construction. If these structures are based on soil layers containing fines, the settlement caused by consolidation should be taken into account when setting the height of the water table in a design cross-section used for stability analysis. (This is because there is a possibility that, after experiencing pore pressure accumulation by an earthquake or any other external loads or events, the borderline might move down and the body of dykes or dams and the dry side might sink down lower than the water table.)

4.6. Surveillance (including periodic inspection and monitoring) and maintenance work on dykes and dams should be performed continually during the construction and operation of the nuclear installation (by a third party or by the dyke or dam operator and safety organization) to prevent and predict potential damage such as the internal erosion of dykes and dams. A safety review of the dykes and dams should be conducted periodically to demonstrate that the dykes and dams are safe, operated safely and maintained in a safe condition and that surveillance is adequate to detect any developing safety problems.

SEAWALLS, BREAKWATERS AND REVETMENTS ON OR NEAR SITES FOR NUCLEAR INSTALLATIONS

4.7. Seawalls, breakwaters and revetments are civil engineering structures used to protect nuclear installations against the wave action of an ocean or a lake during storms and tsunamis. These structures should be properly designed to withstand soil erosion, flooding (including considerations for drainage) and structural failures that might jeopardize items important to safety.

4.8. The effects of waves, tsunamis and earthquakes should be considered in assessing the potential failures of seawalls, breakwaters and revetments. The dynamic effects of waves should be evaluated, including consideration of the maximum static water level derived from flood hazard evaluations, as recommended in IAEA Safety Standards Series No. SSG-18, Meteorological and Hydrological Hazards in Site Evaluation for Nuclear Installations [9].

4.9. The stability of seawalls, breakwaters and revetments should be properly evaluated in relation to the sustainability of their protective functions as well as the effects of their possible failure. The methods of evaluation are similar to those for retaining walls and for the sliding failure of slopes. In performing this evaluation, the material properties of seawalls, breakwaters, revetments and backfill materials, which may include concrete blocks, rubble and other large pieces of material, should be properly estimated. If sandy soils are present at the foot of these structures, their potential for liquefaction should be evaluated and assessed and, if appropriate, resulting consequences should be mitigated.

4.10. Retaining walls can be classified as follows:

(a)   Gravity walls, for which the weight of the wall and possibly of the retained soil play an important part in the wall's stability;
(b)   Embedded walls, such as sheet walls, the stability of which depends on the passive pressure of soil and/or anchors;
(c)   A combination of gravity and embedded walls.

4.11. The input parameters needed to assess the stability of retaining walls are similar to those needed for foundations (see paras 4.25–4.28) and are generally supplemented by geometric data for the soil behind the retaining wall, particularly the slope of the surface. Special care should be taken in determining the level of the water table. Sufficient data should be collected and provided for stability assessment, to a soil depth consistent with the analyses being performed.

4.12. For stability analysis, the pressure of the earth behind the wall may be considered as the active pressure. However, when the admissible displacement of the wall is limited, the pressure of the earth used in analyses and evaluations should be the at-rest pressure.

4.13. For analysis of stability during an earthquake, the inertia forces of the retaining wall and surrounding ground, and the influences of liquefaction or accumulated pore pressure of the ground behind and under the retaining wall, should be taken into consideration. The influences that produce the most unfavourable conditions should be considered in the analysis. If the most unfavourable conditions are not clear, a series of parametric studies of the most extreme conditions for both sides of the wall should be considered. For example, in a pseudo-static evaluation based on seismic coefficients, the vertical component of the seismic acceleration should be considered to be acting upwards or downwards.

4.14. In stability analysis, the failure modes that involve sliding surfaces as well as the failure modes that involve the retaining capacity of the wall should be addressed. The associated safety factors are, respectively, those of the natural slopes and those of the bearing capacities of the foundation. The results of failure mode evaluations might indicate that the movement of a retaining wall becomes larger, and the vertical and lateral displacements of the backfill tend to increase, especially when soil liquefaction occurs in the backfill and/or foundation soil.

FOUNDATIONS OF NUCLEAR INSTALLATIONS

Preliminary foundation work

4.15. Preliminary foundation work comprises the geotechnical activities conducted prior to the placement of the concrete foundations. These activities directly affect the performance of the foundation under the anticipated loading conditions and are therefore essential to safety. They should include the following, as appropriate:

(a) Prototype testing (including test fills and verification of techniques for improving foundation material);
(b) Excavations for foundations or foundation systems;
(c) Dewatering and its control;
(d) Removal of subsurface material (controlled removal techniques should be used to minimize induced fractures below foundations);
(e) Improvement of foundation materials (e.g. modification of material and drainage);
(f) Placement of structural backfill;
(g) Placement of mud mats or any type of protective layer.

4.16. Testing requirements for preliminary foundation work should be specified to ensure proper control and documentation. The testing should include both field and laboratory tests and should be performed throughout the construction stage.

Improvement of foundation conditions

4.17. The phrase 'improvement of foundation conditions' is meant here in its widest sense and includes the modification of the mechanical behaviour of the foundation material (e.g. by soil compaction), the total replacement of loose or soft material by an improved material (consistent with specified quality and performance criteria), or the use of an added material (of sufficient quality) to improve the static and/or dynamic behaviour of the foundations. Another acceptable approach is the use of deep foundations.

4.18. The foundation conditions should be improved if one or more of the following apply:

(a) The foundation material is not capable of carrying the structural loads without unacceptable deformation (i.e. settlements).
(b) There are cavities that can lead to subsidence (see paras 3.4–3.13).

(c)   There are heterogeneities on the scale of the building size that can lead to unacceptable differential settlements.

(d)   The in situ foundation material has shear wave velocities that might lead to unacceptable amplification of the rock input seismic ground motion.

(e)   The in situ foundation material is susceptible to liquefaction.

4.19. When improvement of the foundation conditions is necessary, the following tasks should be performed:

(a)   Characterization of the existing in situ profile and determination of relevant soil parameters pertinent to the selected ground improvement technology;

(b)   Determination of the necessary profile of the foundation material;

(c)   Selection of the particular technology by which improvements in the foundation are to be made (e.g. overexcavation and compacted backfill, rock removal, densification by various methods, solidification by cement or permanent dewatering);

(d)   Performance of a testing programme to experimentally verify the effectiveness of the methods proposed to improve the subsurface conditions;

(e)   Preparation of the specifications for field operations, after the proposed technology has been verified;

(f)   Performance of an investigation at the completion of the improvement programme to determine whether the specifications were met;

(g)   Incorporation of any improvement in foundation material into the design profiles used in the assessments.

4.20. Foundations should not be built on expansive or collapsing soils unless mitigating measures are implemented and it is demonstrated that these phenomena do not adversely impact foundation performance.

Choice of foundation system and construction

4.21. Two foundation systems are available for transmitting the superstructure loads to the soil: shallow foundations and deep foundations. Shallow foundations should be used when the distribution of the load is sufficiently uniform and the upper layers of the soil are sufficiently competent. In the case of weak soil conditions and heterogeneous load, deep foundations should be used to transfer the loads to stiffer soil layers at depth. There may also be separate foundations for separate buildings.

4.22. The following criteria should be applied in the choice of foundation system for a nuclear installation:

(a) The forces due to the structures should be transmitted to the subgrade soil without any unacceptable deformation.
(b) The soil deformations induced by the SL-2 input motion should be compatible with the design requirements of the structure.
(c) The risks associated with the uncertainties in the evaluation of the seismic response should be considered in the design and construction of the foundation system.
(d) The risks associated with underground water should be taken into account.
(e) One single type of foundation should be used for each structure. The choice of the type of foundation should depend on the type of building; for example, a continuous raft should be used under the nuclear island (either supported by piles or founded on competent ground) because it provides homogeneous settlements under static and dynamic loads and can be designed to provide a barrier between the environment and the buildings.

4.23. The analyses and the design profile should represent the behaviour of the structures under the anticipated loading conditions, and therefore the analysis of the foundation systems and structures should represent the as-built conditions.

Analysis and design of the foundation system

4.24. Foundation instability can develop because of inadequate bearing capacity and/or excessive settlements, sliding and overturning; these conditions should be carefully considered, as they can occur because of static or dynamic loadings. Additionally, special consideration should be given to environmental and meteorological conditions and construction activities because they can lead to foundation damage.

Inputs to analysis and design of the foundation system

4.25. Soil and rock characterization should include classification, stiffness and strength, and hydrogeological properties. Engineering properties should include index properties, density, shear strength, seismic wave velocity, elasticity moduli, compressibility, stress state and cyclic resistance. Some of these properties may be strain dependent; testing and reporting of these properties should cover the strain range expected according to the design.

4.26. Rock property characterization should include rock class, type, lithology (e.g. mineralogy, texture), overall geometry (e.g. strike and dip of bedding), discontinuities (e.g. joints, shear zones, fractures), weathering and depositional environment, engineering properties (e.g. mechanical, dynamic, hydraulic, geochemical) and rock mass conditions. The characterization can be performed by means of field and laboratory measurements.

4.27. If the subsurface materials are soils or soft rock, information on the stress history of the subsurface materials should be obtained to predict settlement and heaves and to assess the hazard of gross foundation (shear) failure. Additionally, the creep under static loading of soft rocks (e.g. gypsum, chalk) and clay soil in saturated conditions should be assessed. To compute this stress history, at a minimum the following should be obtained:

(a) The geological stress history and the resulting pre-consolidation stress and overconsolidation ratio.

(b) The loading–unloading history in operations such as dewatering, excavation, backfilling and building construction, as well as the geometry of the disturbed spaces.

(c) The parameters for the establishment and application of the constitutive law applicable to the subsurface materials and their variation with depth, including consolidation parameters. These parameters include the following:
(i) Natural water content;
(ii) Void ratio;
(iii) Liquid limit and plasticity index;
(iv) Compression and recompression indices;
(v) Coefficient of secondary consolidation.

4.28. The following information should be available in the design profile to perform dynamic analyses of the soil–structure interaction:

(a) The best estimate value for body wave (compression and shear) velocity profiles, with a range of variation as determined by in situ measurement techniques. These values should be consistent with the strain levels anticipated from the design basis ground motions.

(b) The number and thickness of layers above the viscoelastic half space. Layering is selected in such a way that each layer has uniform characteristics (i.e. the same soil type and the same shear wave velocity).

(c) The initial conditions of the subsurface materials represented by the shear wave velocity (or shear modulus) at small strain and by Poisson's ratio. These values are determined for each foundation layer of the model.

(d)   The non-linear soil behaviour, which should be taken into account by making use of equivalent linear or non-linear material properties. The design parameters for the equivalent linear method are the shear modulus and the damping versus shear strain relationships for each of the subsurface layers.

(e)   The groundwater level to be used in performing an analysis.

Soil–structure interaction

4.29. Soil–structure interaction is the dynamic interplay between a structure and the soil or rock beneath it during an earthquake or when subjected to dynamic forces applied directly to the structure. The structure's response is influenced by the nature of the vibratory ground motion, the applied static and dynamic loads, the properties of the surrounding soil, and the structure's own characteristics.

4.30. For structures built on hard rock or soils that are very stiff, the motion at the foundation during an earthquake closely matches the free field ground motion at that level. However, for softer rock or soil conditions, soil–structure interaction effects should be evaluated, because the foundation motion deviates from the free field motion.

4.31. Two general approaches exist for analysing soil–structure interaction:

(a)   The direct method, which evaluates the entire soil structure system together in one step, without relying on the principle of superposition. This method operates in both time and frequency domains and can be applied using either linear or non-linear time series techniques.

(b)   The substructuring method, which breaks the problem into parts and solves it in the frequency domain, explicitly by using superposition. It handles the time dependent seismic motion through Fourier transformation methods applied to the input. This approach is limited to linear analysis.

Both methods are valid as long as the physical properties of the foundation medium and structure are appropriately represented.

4.32. Soil–structure interaction effects should be considered for all nuclear structures important to safety, as follows:

(a)   For structures on rock or rock-like materials, these effects may be negligible depending on the amplitude and frequency content of the vibratory ground motion, the structure's natural frequencies and the stiffness of the supporting rock. In such cases, a fixed base model may suffice for seismic analysis.

(b)   In general, soil–structure interaction analysis should be performed for sites with conditions of Type 2 or Type 3 foundation material (see para. 2.42). For Type 1 sites, a fixed base support may be assumed in modelling structures for seismic response analysis.[7]

4.33. The objective of the analysis of dynamic soil–structure interaction should be to calculate the dynamic response of the structure, with account taken of the effects of the coupling between the structure and the supporting foundation medium, when the combined system is subjected to externally applied dynamic loads or earthquake related ground motions.

4.34. Soil–structure interaction analyses should investigate the following:

(a)   The effects of the foundation soil condition on the dynamic response of the structure;
(b)   The effects of buried structures (e.g. scattering effects);
(c)   The effects of dynamic pressure and deformations on the buried structures;
(d)   The global stability and potential uplift and sliding of the foundation;
(e)   The effects of the interactions between adjacent structures through the soil-transmitted couplings.

4.35. The foundation should be designed to resist the forces developed and accommodate the movements imparted to the structure by the design ground motions or the dynamic loading applied to the structure. The dynamic nature of the forces, the expected ground motion, the design basis for the strength and energy dissipation capacity of the structure, and the dynamic properties of the soil should be included in the determination of the foundation design criteria.

Direct method of soil–structure interaction analysis

4.36. In general, soil–structure interaction analysis by the direct method should consist of the following steps:

(1)   Develop a model of the structure.
(2)   Model the foundation including its shape, stiffness and contact with the soil.
(3)   Model the soil by:
       (i)   Defining the soil properties (linear or non-linear);

---

[7] Some States have additional requirements for treating Type 1 sites as fixed base, such as demonstrating that the combination of seismic input, rock properties and structural traits supports a fixed base assumption.

(ii)   Dividing the soil into discrete elements;

(iii)   Positioning the bottom and side boundaries of the model to minimize their impact on the structure's response.

(4)   Define the input motion applied at the boundaries, compatible with the site response analyses.

(5)   Conduct soil–structure interaction analyses.

(6)   Perform follow-up analyses for detailed structural response, if needed.

4.37. The location and type of lateral and bottom boundaries should be selected so as not to significantly affect the structural response at points of interest. Soil discretization (i.e. elements or zones) should be established to adequately reproduce static and dynamic effects.

Substructuring methods of soil–structure interaction analysis

4.38. The substructuring method includes four variants depending on how the degrees of freedom of the soil–structure interface are treated [10]:

(a)   The fixed boundary method, where the interface is assumed to be rigid.

(b)   The flexible boundary method.

(c)   The flexible volume method.

(d)   The substructure subtraction method. Technical justifications should be provided to demonstrate the adequacy of soil–structure interaction analysis based on the substructure subtraction method.

4.39. The four substructuring variants listed in para. 4.38 address the following soil–structure interaction subproblems:

(a)   The site response analysis (applicable to all four variants).

(b)   The structure model (applicable to all four variants).

(c)   The scattering analysis (referring to the inertial and kinematic interaction components), with:

(i)   The fixed boundary method deriving foundation input motion by imposing rigid body constraints on the free field motion developed in subparagraph (a).

(ii)   The flexible boundary method integrating foundation input motion within the full soil–structure interaction analysis, not as a separate output.

(iii) The simplified soil spring method assuming the foundation input matches the free field ground motion [10].

(d)   Foundation impedance (referring to inertial and kinematic interaction):

(i)   The fixed boundary method may use continuum mechanics, tables of data, finite element methods or other methods, typically yielding complex-valued, frequency dependent impedances.

(ii)   The simplified soil spring method often employs frequency independent springs and dampers. Care should be taken to ensure that the non-linear behaviour of the impedances (i.e. stiffness and damping components) is properly reproduced in the soil spring method.

4.40. Similar to the direct method of soil–structure interaction analysis (see paras 4.36 and 4.37), soil discretization (i.e. elements or zones) should be established to adequately reproduce static and dynamic effects. For structures subjected to externally applied dynamic loads, such as wind, blast or forced excitation of vibration, the determination of the dynamic response of the soil–structure system includes the following three basic steps:

(1)   Determining the dynamic properties of the structure (the structural modelling step);

(2)   Determining the force displacement relationships for the foundation medium (the foundation impedance step);

(3)   Determining the dynamic response of the coupled soil–structure system to the applied load (the analysis of the interaction response step).

4.41. The effects on the analyses of uncertainties in the design profile parameters for the foundation material should be considered. These effects should produce a bounding range of results that would envelop the response of the soil–structure interaction system, accounting for the uncertainties. An approach similar to that described in para. 2.50 should be used.

4.42. The foundation soil and the structures exhibit three dimensional dynamic characteristics; consequently, the soil–structure interaction analysis should be conducted in three dimensions to accurately reflect the characteristics of both the soil and the nuclear installation structures.

4.43. The contributions of different types of damping (e.g. material damping, hysteretic damping, radiation damping) should be considered. For soil–structure systems that consist of components (e.g. foundation system, structures, substructures) with different damping characteristics, modelling may use composite modal damping. Maximum limits of damping values should generally be used, but this depends on the models and methods of analysis selected.

4.44. Embedment effects should be considered in the soil–structure interaction analysis of structures supported by embedded foundations. The potential for reduced lateral soil support of the structure should be considered when accounting for embedment effects. A combination of two or more of the following methods can be used to address the partial soil–wall separation:

(a) Estimating the dynamic and static soil pressures to evaluate the separation extent and then adjusting the soil–structure interaction analysis for reduced contact area or soil stiffness near walls.

(b) Assuming no lateral soil connection over the upper half of the embedment or 6 m, whichever is less. Full connection between the structure and the lateral soil elements may be assumed if adjacent structures founded at a higher elevation produce a surcharge equivalent to at least 6 m of soil.

(c) Including the potential for separation and stiffness degradation in the constitutive model of the soils surrounding the foundation and their interfaces.

4.45. Structure–soil–structure interaction is a three dimensional phenomenon that involves dynamic coupling between nearby structures through the soil, where the vibrations of one structure may influence another. Structure–soil–structure interaction can often be ignored for overall structural response, except in cases such as:

(a) For the seismic analysis of a light structure near a massive structure;
(b) For local effects, such as increased wall pressure from an adjacent structure.

In these cases, the effects of structure–soil–structure interaction should be modelled by including all structures in a single soil–structure interaction analysis or by adjusting the input motion at the base of the lighter structure to reflect the influence of the heavier structure on translational and rotational motion.

4.46. Simplifications in structure–soil–structure interaction analysis should be carefully considered before being implemented. The assumption of vertically propagating shear and compressional waves is generally acceptable — except at sites with significant inclined waves or surface waves due to soil conditions — provided that torsional effects from wave angles, rotational motion, and mass or stiffness variations are included through the use of accidental torsion. A loading contribution due to accidental torsion may be included to take into account torsional effects.

4.47. The effect of the incoherency of seismic waves arising from horizontal and vertical variations in ground motions should be considered in the soil–structure interaction analysis. The incoherency stems from:

(a) Random spatial variations due to soil or rock heterogeneity scattering the waves;
(b) Wave passage effects from differing arrival times across the foundation.

Incoherency typically reduces translational foundation motions and increases rotational motions, with greater effects at higher frequencies and larger foundations. Coherency models reflecting the spatial variation with frequency and distance, and soil–structure interaction formulations using these models, should be adequately justified.

Probabilistic analysis of soil–structure interaction

4.48. Where safety objectives and performance goals are defined probabilistically, probabilistic soil–structure interaction analysis may be used to determine the probability distribution of the structure's response and show that the design meets the acceptance criteria (see also paras 4.23 and 4.24 of SSR-1 [1]).

4.49. Probabilistic soil–structure interaction analysis should be performed with simulation approaches. The correlation between simulated parameters should be incorporated into the probabilistic models. A Monte Carlo approach can be used for systems that contain significant non-linear behaviour. For systems that have essentially linear behaviour or that include minor non-linear responses, Monte Carlo or a more efficient stratified sampling approach, such as a Latin hypercube simulation, may be used by treating key seismic response parameters as random variables [10].

4.50. Inputs for probabilistic soil–structure interaction response analysis should include a number of ground motion sets, represented as acceleration time series or response spectra, each with two horizontal components and one vertical component.

4.51. A set of response analysis simulations should be conducted by sampling random values from identified parameters for each run. Then, the statistical properties of selected responses should be assessed. Given the computational demands of soil–structure interaction response analysis, a Latin hypercube simulation is generally used.

4.52. Seismically induced soil–structure interaction effects related to foundation overturning and sliding should be considered, as should potential differential displacement for single foundations and between piping and conduits that are important to safety and are connected to the foundation or the superstructure.

Contact pressure beneath foundations

4.53. The distribution of contact pressure beneath the foundations and the stresses induced in the subsurface materials should be derived from the analysis of the static soil–structure interaction. In addition to the elastic and geometric parameters of the structures (e.g. geometry and stiffness of the foundation mats and of the superstructure of the buildings), the mechanical characteristics of the subsurface materials should be included in the design profile to allow the foundation contact pressure to be computed.

4.54. The most widely used type of foundation for nuclear power plants is the mat type (foundations other than mats are commonly used in other types of nuclear installation; see para. 4.22). The design of the foundation mat should be analysed for relevant types of structural stiffness behaviour (e.g. infinitely rigid foundation, flexible foundation, actual structural stiffness). The stiffness of the superstructure should be taken into account if it is needed in the analysis. To compute the distribution of contact pressure under the foundation, the subsurface foundation material can be modelled by the finite element method (i.e. continuum representation) or by representing the foundation as a series of springs whose stiffness corresponds to the coefficient of subgrade reaction (i.e. lumped representation).

4.55. For the two extreme conditions of infinitely stiff and infinitely flexible foundations (in the case of distributed load on soil), general solutions are available in foundation design textbooks and design standards. For intermediate conditions, which occur most commonly, numerical solutions using computer codes are usually employed. Consideration should be given to the condition in which the stiffness of the structures changes as construction proceeds. Any non-linear behaviour exhibited by the subsurface materials when subjected to unloading and reloading during excavation, dewatering and backfilling should also be considered.

4.56. For structures located close together, the possible impacts of adjacent structures on the response of the foundation soil should be evaluated. In this case, a three dimensional analysis should be considered.

4.57. The analysis of foundation stability should be performed under static (i.e. permanent) loads and under a combination of static loads and dynamic loads induced by seismic acceleration. The vertical component of the seismic ground motion should be considered to be acting upwards or downwards. The assessment should consider bearing capacity, overturning and sliding.

4.58. The cyclic seismic forces generated in the foundation material by the seismic ground motion should be computed by an appropriate dynamic method to derive the maximum of these forces. These forces can be converted to equivalent static forces for the assessment of stability. The same dynamic method may be applied to the analysis of uplifting and overturning and to the computation of lateral loads on subsurface walls and retaining walls. The equivalent static forces should be derived in accordance with the stability analysis under consideration. The use of a non-linear or linear time history approach to show stability for seismic loading should be considered.

4.59. In the case of an embedded foundation, active pressure of the soil should be regarded as an additional horizontal load.

4.60. For structures founded above the groundwater level, the angle of shearing resistance between soil and structure should be less than or equal to the angle of effective shearing resistance for cast-in-place foundations and should be less than or equal to two thirds of the angle of effective shearing resistance for precast foundations.

4.61. If the sliding resistance is the sum of shear friction along the foundation and the soil lateral pressure (up to the full passive pressure capacity induced by embedment effects), a consistent lateral displacement criterion for activating the passive soil pressure should be used. This involves the use of a static (as opposed to dynamic) coefficient of friction, consistent with the use of partial versus full passive pressure.

4.62. The sliding safety analysis of the foundation of a nuclear installation should include not only an assessment of the balance of forces between the resistance and the design load but also a comparison of the displacements (evaluated by appropriate methods, such as the finite element method or the boundary element method) during and after the seismic ground motion with the acceptable value.

4.63. For static loading, analysis of stability against sliding and overturning should provide adequate factors of safety against sliding and overturning.[8] The analysis should consider variations in loading during the lifetime of the structure due to such factors as rise in groundwater level, removal or reduction in passive forces downslope (for any reason), increase in driving forces upslope (for any reason), liquefaction potential, or other factors.

4.64. For analysis of stability against overturning, a ground contact ratio — defined as the ratio of the minimum area of the foundation in contact with the soil to the total area of the foundation — may be used. The seismic response computed over the entire duration of the seismic ground motion should be considered to determine the minimum value of this ratio. If the defined minimum contact area[9] is not achieved, then the non-linearity due to the foundation uplift should be assessed and, if found to be important, should be accounted for in the design.

4.65. Under certain combinations of ground motion, groundwater level and geometrical configuration of the building, conventional computing procedures might suggest a potential uplift. This does not mean that the foundation will necessarily lift up, but rather that conventional procedures to compute the structural response might not be applicable under these circumstances. If the estimated surface area of the uplift of the foundation is larger than the defined contact area limit (as a percentage of the total surface of the foundation), a more sophisticated method should be used in the analysis of the dynamic soil–structure interaction. The estimated uplift of the foundation should be limited to a value that is acceptable in respect of the bearing capacity of the soil and the functional requirements.

4.66. The analysis of foundations should consider the effects of the bending moment and shear forces in the foundation induced by static and dynamic loads, the buoyant load, the potential foundation lift-off effect and embedment effect, as well as the effect of various sliding interfaces on selections of coefficients of friction (e.g. soil shear failure, concrete to soil, waterproofing to soil, concrete basemat to concrete mudmat).

---

[8] Some States define the minimum factor of safety against sliding and overturning under dynamic loadings as 1.1; other States define the minimum factor as 1.5. The acceptable safety factor depends on the method of analysis, the definition of capacities and other considerations.

[9] Differing definitions of minimum contact areas exist. Some States set a minimum value for the ground contact area ratio as low as 70%, corresponding to a 30% uplift; other States set the minimum value as high as 80% for overturning and 20% for uplift.

4.67. Uncertainties in dynamic foundation sliding and rocking responses should take into account variable friction coefficients, cohesion strength and related parameters to estimate behaviour and ensure the design meets the acceptance criteria of the regulatory body.

Settlement and bearing capacity

4.68. Foundations should be analysed to ensure adequate bearing and tolerable settlement of the underlying soils. The analysis should include assessment of geological materials extending to a sufficient depth within the zone of influence of foundations. The analysis should consider uncertainties due to materials, models and loads.

4.69. Linear and/or non-linear methods may be used for settlement analysis. Both total settlement and differential settlement due to elastic compression, consolidation, secondary compression and dynamic settlement over the lifetime of the nuclear installation should be considered.

4.70. An analysis of settlement under static loads should be performed. The possibility of differential settlements or heaves between the buildings of a nuclear installation potentially affecting connecting pipes, conduits and tunnels should be investigated. Settlements and heaves are also important in relation to deformation of the foundation, which could lead to overstressing of buildings and interference with the operation of machinery such as pumps and turbines if this machinery is not isolated from its supports.

4.71. Short and long term settlements occurring during the operating lifetime of the nuclear installation should be estimated.

4.72. Time dependent settlements may be computed by applying the classical theory of consolidation or other sophisticated non-linear analyses. In saturated soils, the following three components should be considered:

(a)   Undrained shear settlement;
(b)   Settlement caused by consolidation;
(c)   Settlement caused by creep.

4.73. The following actions should be taken to analyse long term settlement:

(a) The anticipated loading history of the subsurface materials should be specified (e.g. excavation sequence, dewatering process, backfilling, construction process).

(b) For each layer, a model should be chosen in accordance with data from laboratory and in situ testing.

(c) The models should be assessed and improved by interpreting the measurements for settlement and heave made during excavation, dewatering, backfilling and construction.

(d) The models should be corrected by comparing their predictions with observations so that any necessary adjustments can be introduced for their use in future predictions.

4.74. A conservative analysis of differential and total settlement should be performed for the design of the foundations for buildings, interconnecting structures between adjacent buildings and foundations for machinery.

4.75. Seismically induced settlements should be considered in settlement analysis. Settlement effects from other potential vibratory sources should also be included in these evaluations, if appropriate. Other effects causing additional settlements (e.g. changes in ground elevation, adjacent excavations, hydrogeological conditions) should also be considered in the settlement analysis, as appropriate.

4.76. If no structure–soil–structure interaction analysis was performed, a soil–structure interaction analysis should be performed structure by structure, and the individual displacements should be combined to obtain the dynamic part of the differential displacement. Both horizontal and vertical components and their combinations should be considered.

4.77. The effects of the construction sequence and the effects of the installation of systems and components on settlements should be assessed.

4.78. For structures located on soils that might exhibit permanent seismically induced vertical or lateral deformations, the effects of the permanent deformations should be evaluated.

4.79. The method for computing the ultimate bearing capacity should be consistent with the assumptions associated with the soil conditions and the chosen approach. Classic soil mechanics methodologies for computing the ultimate load bearing capacity are acceptable if the subsurface material is relatively uniform.

Analysis of elastic plastic equilibrium can be performed for the plane strain and the axially symmetric cases. The main difficulty in this analysis is the selection of a mathematical model of soil behaviour or its constitutive (stress–strain–time) relationship. The available solutions are generally limited to those developed for the rigid plastic solid. This solid is assumed to exhibit no deformation prior to shear failure and a plastic flow at constant stress after failure.

4.80. If the subsurface material exhibits considerable heterogeneity, anisotropy or discontinuity, the sliding surface method should be used instead of the bearing capacity formulas. In this method, potential sliding surfaces with smaller safety factors for sliding are predetermined for the subsurface material and analysed in a conventional slip surface analysis for behaviour under the initial static load and equivalent seismic load. If the calculated safety factor is lower than acceptable, further analysis should be performed. A dynamic analysis using acceleration time histories under the initial static load may be performed. In all these analyses, the vertical seismic force should be taken into account in a conservative manner.

4.81. For cohesive soils, both short term and long term bearing capacities should be assessed.

4.82. Estimates of ultimate capacity should include dynamic effects and should not be based on standard relationships associated with general shear failure concepts appropriate for static load cases.

4.83. For cohesive soils or saturated cohesionless soils, earthquake induced strength degradation (associated with cyclic softening or excess pore water pressure generation) should be used in bearing capacity evaluations.

4.84. The water level should be assumed to be equal to the highest water level expected due to the design basis flood for static loading. The groundwater level should be assumed to be the mean level for the determination of the bearing capacity under seismic loading.

4.85. The cyclic seismic forces generated in the foundation material by the seismic ground motions should be computed by an appropriate dynamic method to derive the maximum of these forces and to estimate the number of equivalent loading cycles, if this parameter is also necessary for the assessment of bearing capacity.

4.86. The potential for failure of the bearing capacity of the subsurface materials for a nuclear installation under static loading should be low, so that there are high margins of safety under static loading (this is generally the case). These margins should be sufficient to meet seismic loading conditions with reasonable safety margins.

4.87. If an adequate safety factor is achieved on the basis of conservative assumptions, no further analysis is generally necessary. Acceptable safety factors depend on the method of analysis and on other considerations. In the conventional method, the safety factor should be consistent with national and/or international codes and standards, including combinations of loads that involve seismic input (the overturning effect). Reliability analysis, including load and resistance factor design approaches, may be used to demonstrate that an adequate margin is included in the design.

4.88. Where fractured rock is present as foundation material, a local safety factor should also be included. The local safety factor is defined as the ratio of the strength to the working stress at each point where yielding or local sliding along the existing fracture zones and weathered zones beneath the foundation might occur. This factor indicates the extent of the yielding zones or the progressive failure of the material subjected to the design load. It is useful in determining the position and extent of the improvements that may be needed in foundation materials and in choosing an appropriate technique for the improvements. If, under combinations of loads that involve the seismic input, this safety factor is lower than 1 in an area sufficiently large that it would affect the performance of the structure, foundation conditions should be improved. However, the macroscopic stability should be judged on the safety factors for bearing capacity and sliding.

Heave effects on foundations

4.89. The effects of frost depth and frost heave should be considered in the analysis of shallow foundations.

4.90. In areas subjected to frost heave, spread footings and mats should be placed below frost depth or designed to have sufficient uplift resistance to overcome forces due to ground heave and frost jacking[10]. The structural design of foundation connections should be sufficient to transmit the loads due to frost heave and adfreeze[11].

---

[10] Frost jacking is the frost heave process that involves upward displacement of an object embedded in freezing soil.

[11] Adfreeze is the process by which two objects are bonded together by ice formed between them.

4.91. Where shallow foundations are placed above the seasonal frost depth, they should be protected from frost heave effects using frost-protected shallow foundations[12].

4.92. The effects of heave due to excavation and unloading, expansive soils or rocks, and glacial rebound should be evaluated where applicable.

EARTH STRUCTURES ON SITES FOR NUCLEAR INSTALLATIONS

4.93. The design of earth structures (e.g. retaining walls, dykes, levees, duct banks) important to safety at a nuclear installation should be consistent with the design of the installation itself. In particular, the external hazards against which those structures are designed should be consistent with the events selected in the design of the nuclear installation; these events and their associated loads should be listed in the contractual terms of reference relating to the earth structures. The list should be supplemented by the specific events that could challenge the safety of these structures.

4.94. The time, extent and duration of seismic aftershocks are unpredictable. Consequently, changes in soil states after a main shock should be taken into account for aftershock safety assessments and evaluations. For example, degradation of soil rigidity and strength might result from decreased confining pressure caused by excess pore water pressure, which could take considerable time to dissipate.

4.95. At sites that are expected to experience inundation caused by a flood or tsunami, potential ground erosion — including changes in geometry and material properties — should also be taken into account for evaluations, bearing in mind the nature of the event (e.g. duration, peak flow, maximum water height). This is particularly important when considering phenomena related to water flows that lead to the failure of earth structures or soil foundations, such as internal and external erosion and scouring.

4.96. Evaluations of the consequences of the failure of earth structures that are important to safety should be conducted, with particular consideration of the significance and purpose of these structures.

---

[12] A frost-protected shallow foundation is a foundation that does not extend below the design frost depth but is protected against the effects of frost using, for example, expanded polystyrene and extruded polystyrene.

4.97. The consequences of the failure of earth structures that are important to safety, and any structures, systems and components dependent on them that could endanger items important to safety, should be evaluated against stability and/or deformation criteria.

4.98. The consequences of the failure of earth structures that are only indirectly related to the safety of the nuclear installation (i.e. that are not important to safety but that could have an impact on the site or on structures, systems or components important to safety) should also be taken into consideration. To simplify evaluations of complex interactions with such structures, stability analyses can be conservatively adapted, provided that the consequences remain insignificant.

EMBEDDED STRUCTURES ON SITES FOR NUCLEAR INSTALLATIONS

4.99. Embedded structures are buildings with foundations deep enough that the interaction of the underground walls with the surrounding ground is significant. Two aspects of such embedment should be considered, as follows:

(a)   Underground walls acting as retaining walls (see paras 4.10–4.14);
(b)   The effects of the soil on the structure (see paras 4.100–4.103).

4.100.   The input parameters for the analysis of embedded structures are similar to those for foundations and retaining walls, and information on them should be obtained accordingly. Supplementary information should be obtained on the safety and serviceability criteria for the underground walls to be met under different loading cases (particularly in relation to leaktightness). For this purpose, the possible cracking of concrete (i.e. limiting the stresses in reinforcement bars and concrete) should be taken into account in the design of the foundation and the construction joints of buildings. For embedded structures credited as or considered to be containment structures, recommendations on containment considerations are provided in IAEA Safety Standards Series No. SSG-53, Design of the Reactor Containment and Associated Systems for Nuclear Power Plants [11].

4.101.   The challenging effects of groundwater on both the stability and the leaktightness of embedded structures should be taken into account in the design. Drainage should be incorporated for any foundation beneath the level of the water table, or the hydrostatic pressure should be taken into consideration. At coastal sites, the possible adverse effects of varying levels of groundwater salinity on the foundation material and isolation material should be considered.

4.102. A building can be regarded as embedded only if the backfill has been properly compacted or if other appropriate measures have been taken. In such cases, the effects of embedment on the impedance of the foundation and on the soil–structure interaction should be taken into account. If the building is not effectively embedded, only the consequences of the depth of the foundation should be taken into account, disregarding the effects of the interaction of soil with the underground walls.

4.103. For stability analysis of effectively embedded foundations under seismic loads (see paras 4.57–4.67), the friction between soil and walls should be disregarded.

BURIED PIPES, CONDUITS AND TUNNELS ON SITES FOR NUCLEAR INSTALLATIONS

4.104. The layout of buried pipes or conduits should be considered in the geotechnical site investigation programme. Adequately spaced boreholes, drillings, soundings and/or test pits should be made along the pipe routes. Special consideration should be given to identifying areas of discontinuities or changes in the foundation material along the route of the piping. The routeing of pipes or cables via areas susceptible to inundation by floods or tsunamis should be avoided. This applies also for buried pipes or conduits, if feasible. In areas susceptible to frost, the effects of frost depth and frost heave should be considered in the design and analysis of buried pipes and, if necessary, frost protection measures should be implemented.

4.105. The depth of investigation boreholes, drillings, soundings or test pits depends on the stratigraphy of the foundation material below buried piping but, at a minimum, should extend to a competent soil layer below the foundation level.

4.106. An assessment of the potential effects of any corrosive environmental agents on the piping material should be included in the site investigation programme.

4.107. Buried piping should be placed at a sufficient depth to prevent damage due to surface loading (e.g. traffic loads) or should be designed to resist the surface loads to which the pipes will potentially be exposed. Buried piping should also be placed at a sufficient depth to prevent damage or non-functionality due to freezing.

4.108. Piping should be placed on well compacted granular material over competent foundation material, so that no damage or distortion of the piping due to displacements (e.g. heaving, settlement, lateral spreading) or liquefaction of the foundation material can occur. Foundation improvement techniques may be used for weak subsurface conditions.

4.109. Buried piping, conduit systems and tunnels important to safety should be designed to resist the effects of earthquakes.

4.110. Long, buried piping systems are primarily subjected to relative displacement induced strains rather than inertial effects. These strains are induced primarily by the passage of seismic waves and by differential displacement between a building attachment point (i.e. an anchor point) and the ground surrounding the buried system. The following seismically induced loadings should be considered for long, buried piping and for conduits and tunnels:

(a) Strains induced by the passage of seismic waves;
(b) Differential displacements in zones of different materials;
(c) Additional loads due to seismic oscillations resulting in sloshing of internal liquids;
(d) Deformation and shaking of the ground or anchor points relative to the ground;
(e) Ground failures, such as surface fault rupture, liquefaction, landslides, settlements and discontinuous displacements.

4.111. For deep tunnels and shafts, hoop stresses and strains also develop owing to travelling seismic waves, and these hoop strains should be considered in the design.

4.112. In the analysis of the effects of seismic ground motions on the piping system, the following two types of loading should be considered:

(a) Relative deformations imposed by seismic waves travelling through the surrounding soil or by differential deformations between the soil and anchor points;
(b) Lateral earth pressures acting on the cross-section of the structural element.

4.113. For sections of a long, linear buried pipe that are remote from anchor points, it may be assumed that sharp bends or intersections move with the surrounding soil and that there is no movement of the buried structure relative to the surrounding soil. In this case (unless the movement is otherwise justified), the maximum axial strain can be estimated by ignoring friction between the piping and the surrounding soil. If there is a possibility of slippage between the pipe and the surrounding soil, the axial strain for straight sections remote from anchor points, sharp bends or intersections should be estimated, with account taken of the friction. An estimate of axial strain depends on the wave type that results in the maximum ground differential displacements. The wave types that should be considered are compression waves, shear waves and surface waves.

4.114.   In addition to the forces and strains in the buried pipes due to wave propagation effects, the forces and strains due to the maximum relative movement between anchor points (e.g. building attachment points) and the adjacent soil, which occurs as a result of the dynamic response of the anchor point, should be calculated. In calculating maximum forces and strains in the buried piping, the motion of adjacent anchor points should be considered in a conservative manner.

4.115.   Factors that could compromise the function of buried pipes, conduits or tunnels, such as discontinuous displacements (both parallel and perpendicular to the length of the system), axial strains and inclinations of the structure, should be evaluated.

4.116.   In the analysis of tunnels, the stresses and deformations due to all expected loads, including earthquake motions, should be considered. Stresses can be assessed empirically or numerically, such as by the finite element method.

4.117.   The consequences of the failure of ducts and pipes and other underground features passing near or through structures important to safety at the nuclear installation site should be given appropriate consideration. If hazardous effects are expected, appropriate measures should be taken to protect the installation; alternatively, the site layout should be reconsidered.

5. MONITORING OF GEOTECHNICAL PARAMETERS ON SITES FOR NUCLEAR INSTALLATIONS

PURPOSE OF MONITORING GEOTECHNICAL PARAMETERS ON SITES FOR NUCLEAR INSTALLATIONS

5.1.   A documented site monitoring programme should be established, at the latest when a site is selected. Depending on the monitoring objectives, a baseline monitoring programme may be needed to document the undisturbed environment and provide data for the preservation of evidence. The monitoring programme should identify and address the need for specific data (including monitoring frequencies and management practices), methods of monitoring, and overall interpretation and review expectations. The monitoring plan should be evaluated regularly, but the baseline monitoring data should be preserved to enable the comparison of safety relevant parameters with the baseline values.

5.2. Field monitoring, in particular quantitative measurements of performance outputs, should be implemented to define and monitor the geotechnical parameters necessary for the safe design, construction and operation of the nuclear installation. Electrical devices have become the standard method of monitoring and are widely used in geo-monitoring applications.

5.3. Subsurface investigations, in situ testing and laboratory testing should be undertaken to provide values of parameters and information on site characteristics suitable for predicting the performance of foundation systems under the envisaged loading conditions. The use of these parameters enables criteria for the foundation design to be established for the performance of the foundation materials and structures under anticipated loadings. To verify the performance of the foundations and earth structures, their actual field behaviour should be monitored from the beginning of siting activities, through construction, to the end of the operation of the nuclear installation.

5.4. The monitoring of actual loads and deformations enables a field check to be made of the predicted behaviour of the foundations and buried structures. Since the construction stage is generally lengthy, the monitoring data allow the settlement models to be revised on the basis of actual performance. Predictions of long term performance can therefore be made with reasonable confidence.

GUIDELINES FOR MONITORING SITES FOR NUCLEAR INSTALLATIONS

5.5. The stages of construction of a nuclear installation usually consist of excavation, backfilling and building construction. The behaviour of the soil should be monitored during each of these stages. During the excavation and backfilling stages, deformation of subsurface material (e.g. heave and settlement, lateral displacements) should be monitored, and load evaluations should be made. Monitoring should be continued throughout the lifetime of the nuclear installation.

5.6. The groundwater conditions under buildings and in adjoining areas at the site of a nuclear installation should be monitored to verify the conditions outlined in the design assumptions, especially if deep drainage systems or permanent dewatering systems are installed. Groundwater monitoring should be undertaken early in the geotechnical investigation to inform the hydrological and hydrogeological models and should be continued throughout the lifetime of the nuclear installation.

5.7. Deflection, displacements and relevant parameters of structures important to safety, including retaining structures and earth structures, should be monitored.

5.8. The seismic behaviour of the nuclear installation site and the subsurface materials should be monitored. The need for instrumentation[13] to monitor the in situ pressure of pore water for liquefaction studies should also be considered.

5.9. If the site has the potential for slow bedrock movements, such potential relative movements between recognized bedrock blocks (e.g. on opposing sides of fracture zones) should be monitored.

5.10. Monitoring devices should be carefully chosen so that the monitoring system provides the expected information over the lifetime of the installation. The choice of devices should be informed by feedback on the experience of monitoring other sites for nuclear installations. In deciding on the number of devices and manual measurement points, the expected failure rate of these devices should be evaluated, with special consideration of their need for replacement.

5.11. If a specific geotechnical site monitoring device needs to be replaced, the replacement procedure should be documented in detail. The new device may represent an updated technology, and direct equivalence in measurement capacity is not compulsory, provided that the minimum specifications for resolution, accuracy, data collection and environmental impact during installation are satisfied. Where possible, a final set of measurements should be taken from the device to be replaced, to ensure the calibration of the measurements from the old device with respect to reference measurements from the new device.

5.12. The geotechnical site monitoring programme should be documented, clearly indicating the procedures for the collection, standardized storage, management and visualization of the data. The programme should include the necessary qualifications of technical personnel, the specification and qualification of hardware and software systems that collect and report data, and the protocols

---

[13] An example of such instrumentation is pore water pressure transducers (piezometers), which are capable of measuring dynamic changes in excess pore water pressure.

for data dissemination. The monitoring programme and monitoring records should include the entire monitoring history of the nuclear installation, from site selection, through the construction, commissioning and operation stages to decommissioning.

5.13. A periodic review of the monitoring programme should be performed. The review period should be dependent on the results of the monitoring itself, the rate of technological advances in the field, geotechnical and/or structural requirements during the lifetime of the installation, and any other conditions that would necessitate an updated monitoring programme.

MONITORING DEVICES FOR SITES FOR NUCLEAR INSTALLATIONS

5.14. Specifications for the selection of geotechnical site monitoring devices — including preferences in terms of sensors, data acquisition systems and related components and accessories — should be defined on the basis of an assessment of long term exposure to environmental conditions (including atmospheric conditions, temperature, hydrogeological conditions, hydrochemical conditions, electromagnetic interference and sources of background noise) and the necessary measuring precision. For recommendations on seismic monitoring devices, see paras 3.54–3.59 of SSG-9 (Rev. 1) [2].

5.15. All operational geotechnical site monitoring devices should be regularly maintained. Procedures for maintaining commissioned monitoring devices should be defined and should be documented in accordance with the management system. These procedures should include, where applicable, protocols for harmonizing the data obtained from failed devices with the reference readings of the newly installed equivalents. Additionally, data harmonization and calibration should be ensured among all operational devices of different types, technologies or methods of measurement (e.g. digital, digital output with manual data collection procedure, fully digitized and automated systems, or fully manual and analogue systems).

5.16. Monitoring devices should be used to observe the behaviour of the foundation and related materials. Table 4 contains a list of available devices that can be used for monitoring soil and buildings (e.g. extensometers, load and pressure cells), depending on the site, the monitoring requirements and the type of nuclear installation.

TABLE 4. EXAMPLES OF GEOTECHNICAL MONITORING DEVICES FOR SITES FOR NUCLEAR INSTALLATIONS

Type of device	Principle	Location	Parameter measured	Purpose
Piezometer, water level meter	Hydraulic pressure	Boreholes, reservoirs, weirs	Pore pressure, water table	Monitoring of water table, positive and negative pore pressure monitoring, hydrogeological characterization, monitoring of water level in reservoirs, drainage channels and weirs
Hydraulic device	Hydraulic, U tube, hydraulic load cells	Basement and beneath, isolated foundations of operating machinery	Deformations and stresses of the basemat, loads on soil nails, rock bolts and prestressed ground tendons	Behaviour of the soil–structure system, high-sensitivity settlement monitoring of foundation systems
In situ settlement plate	Topography	Ground surface	Displacements, settlements	Settlement of structures
Survey reflector	Topography	Ground surface, fill layer base or along intermediate layering within fills	Displacements, settlements	Settlement of structures and fills
Rod extensometer	Mechanical, electromechanical	Boreholes, excavation support structures	Settlement, heave, lateral deformations, stability of jointed rock masses	Deformation of structures, stability of natural soil and rock slopes

TABLE 4. EXAMPLES OF GEOTECHNICAL MONITORING DEVICES FOR SITES FOR NUCLEAR INSTALLATIONS (cont.)

Type of device	Principle	Location	Parameter measured	Purpose
Magnetic extensometer, induction current type extensometer	Electromagnetism	Boreholes	Settlement, heave	Deformation of fills and human-made slopes
Gammagraphy, photogrammetry	Superposition of picture	Ground surface	Deformation of topography	Deformation of structures
Global positioning system	Aiming by satellite	Ground surface, site	Topography of the site, XYZ coordinates (particularly Z)	Site evaluation, relative movements between bedrock and blocks
Interferometric synthetic aperture radar (InSAR)	Synthetic aperture radar	Remote sensing of ground surface	Deformation	Settlement of structures, ground subsidence
Georadar	Radar based proximity measurement	Ground surface	Distance	Deformation of structures, monitoring performance of slopes

TABLE 4. EXAMPLES OF GEOTECHNICAL MONITORING DEVICES FOR SITES FOR NUCLEAR INSTALLATIONS (cont.)

Type of device	Principle	Location	Parameter measured	Purpose
Lasermeter	Laser light source	Ground surface, underground openings, interior spaces in industrial facilities	Distance	Behaviour of structural systems, convergence of underground openings
Inclinometer, tiltmeter, pendulum system	Electromechanical, electrolytic, microelectromechanical systems, optical, laser	Borehole, isolated locations on structural members, embankments, fills, route structures, tall structures	Tilt, absolute inclination, deformation profile derived from tilt measurements along predefined axes, three dimensional deformation profile using three dimensional measurement of inclination along an array	Stability of slopes, embankment loading related deformations, retaining structures, walls, determination of fill settlement profile, performance of machine foundations
Crackmeter, jointmeter, tape extensometer	Electromechanical, mechanical	Surface of structural members, foundation members, retaining structures, surface of rock masses along discontinuities	One dimensional to three dimensional displacement measurements	Performance of structural and architectural joints, construction joints, performance of retaining structures, slope stability monitoring, pre-failure identification of unstable rock masses (e.g. rock fall hazards, toppling, planar and wedge type failures)

TABLE 4. EXAMPLES OF GEOTECHNICAL MONITORING DEVICES FOR SITES FOR NUCLEAR INSTALLATIONS (cont.)

Type of device	Principle	Location	Parameter measured	Purpose
Soil extensometer	Electromechanical	Soil mass (embankments), superstructures	Lateral deformations under tensile stresses	Crack under tensile stresses, lateral movements in embankments and fills
Strain gauge	Electromechanical, fibre optic	Deep foundation elements, deep excavation elements, basemats, tunnel and gallery linings, embedded within soil for the case of distributed fibre optic strain sensing	Strain (uniaxial, biaxial, triaxial)	Behaviour of soil–structure system (e.g. deep foundations, deep excavations), foundation stress distribution, deformation monitoring of rock slopes
Earth pressure cell, stress cell	Electromechanical	Embankments, retaining structures, tunnel and gallery linings	Total earth pressure, stresses within concrete members	Monitoring of vertical and lateral earth pressures, measurement of lateral earth pressure coefficient, monitoring of behaviour of underground openings
Load cell	Electromechanical	Soil nails, rock bolts, prestressed ground anchors	Loads on soil nails, rock bolts and prestressed ground anchors, piles	Behaviour of the soil–structure interaction system, performance verification of piles

TABLE 4. EXAMPLES OF GEOTECHNICAL MONITORING DEVICES FOR SITES FOR NUCLEAR INSTALLATIONS (cont.)

Type of device	Principle	Location	Parameter measured	Purpose
Soil electrical resistivity monitor	Electrical resistance or conductivity	Soil	Electrical resistance or conductivity of soil	Monitor changes in soil conditions and characteristics over time
Seismometer	Accelerometers, triggers	Free field, buildings, boreholes	Acceleration time histories	Operability of nuclear installations, seismic behaviour of structures and sites, floor response spectra, early warning triggers due to natural hazards
Acoustic emission	Acoustic signal emission	Ground surface, underground openings, pipeline systems	Acoustic waveform, time and frequency domain waveform analysis	Detection of leaks in buried piping, early detection of unstable rock masses in slopes and underground openings
Temperature sensing	Thermistors, resistive temperature detectors, thermocouple action, contact based distributed fibre optics	Mass concrete (embedded), concrete, steel (surface), soil mass, embankment, drainage features, boreholes	Temperature, spatial and temporal variation of temperature	Seepage detection, temperature induced strains and stresses, structural integrity (piles), performance of steel structural systems, performance of energy piles, buried pipelines
Tachymeter or tacheometer	Laser	Ground surface	XYZ coordinates (particularly Z)	Relative movements between bedrock and blocks

5.17. Monitoring of structures important to safety should include total and differential settlements, lateral displacements and deformations, earth and pore pressures, and inclinations along sloping ground surfaces. Monitoring of the performance of other structures with a potential impact on items important to safety should also be considered.

6. APPLYING A GRADED APPROACH TO GEOTECHNICAL ASPECTS IN THE SITING AND DESIGN OF NUCLEAR INSTALLATIONS

6.1. For site evaluation of nuclear installations other than nuclear power plants, a graded approach is required to be applied (see Requirement 3 and paras 4.1–4.5 of SSR-1 [1]). In the application of a graded approach, the complexity of the site should be taken into consideration.

6.2. The application of a graded approach to the geotechnical site investigation and characterization (see Requirement 22 of SSR-1 [1]) might increase the uncertainty in the geotechnical parameters used as input for the design bases. This larger uncertainty might result in a reduction of the reliability of the design. It should be ensured that any reduction of reliability is considered acceptable with respect to the overall safety objectives.

6.3. The risk associated with a nuclear installation depends on the potential failures within the installation and on the on-site and off-site consequences of such failures. The overall safety objective in site evaluation, as established by Requirement 1 of SSR-1 [1], is the same for all nuclear installations. However, for a particular nuclear installation, the radiological consequences of failures might be so small that reliability levels lower than those for high radiological hazard facilities could be accepted without compromising the safety objective.

6.4. The radiological consequences of potential failures depend on the nature of the nuclear installation and the characteristics of the site. Paragraph 4.5 of SSR-1 [1] states:

"For site evaluation for nuclear installations other than nuclear power plants, the following shall be taken into consideration in the application of a graded approach:

(a) The amount, type and status of the radioactive inventory at the site (e.g. whether the radioactive material on the site is in solid, liquid and/or gaseous form, and whether the radioactive material is being processed in the nuclear installation or is being stored on the site);

(b) The intrinsic hazards associated with the physical and chemical processes that take place at the nuclear installation;

(c) For research reactors, the thermal power;

(d) The distribution and location of radioactive sources in the nuclear installation;

(e) The configuration and layout of installations designed for experiments, and how these might change in future;

(f) The need for active systems and/or operator actions for the prevention of accidents and for the mitigation of the consequences of accidents;

(g) The potential for on-site and off-site consequences in the event of an accident."

6.5. The application of a graded approach to the geotechnical site investigation should be based on a site specific consequence analysis (simplified, as appropriate) that categorizes the installation in terms of the radiological hazard. Four radiological hazard categories are defined in Table 5: 'high', which corresponds to large nuclear power plants; 'medium', which corresponds to installations with potential for significant on-site releases; 'low', which corresponds to facilities that only have potential for localized releases; and 'conventional', which corresponds to conventional industrial facilities, with a negligible (or no) radiological hazard.

6.6. The simplest consequence analysis that should be performed corresponds to an unmitigated release of the full radioactive inventory present in the nuclear installation. This is a conservative bounding analysis and provides a first approximation of the hazard category of the nuclear installation. If the result of

TABLE 5. CATEGORIES OF NUCLEAR INSTALLATIONS BASED
ON POTENTIAL RADIOLOGICAL CONSEQUENCES RELEVANT TO
GEOTECHNICAL ASPECTS

Hazard category	Consequences on the site	Consequences off the site	Remarks
High	Radiological or other exposures that might cause loss of life of workers in the facility	Potential for significant off-site radiological consequences	Graded approach is not applicable
Medium	— Potential for significant on-site consequences — Unmitigated radiological release would necessitate site evacuation	Small potential for off-site radiological consequences	See para. 6.10
Low	Potential for only localized radiological consequences (within 30–100 m of the point of release)	No off-site radiological consequences	See para. 6.10
Conventional	No radiological consequences	No radiological consequences	Geotechnical investigation with the same scope as for conventional industrial facilities

such a radioactive release is negligible radiological consequences (for workers, the public and the environment), then the installation should be classified at the lowest radiological hazard category and the geotechnical design basis should be established in the same way as for a conventional industrial facility.

6.7. The results of consequence analyses, in which a design-dependent set of source terms is used and credit is taken for some engineered mitigating features, should be considered acceptable for the radiological hazard categorization of a nuclear installation, provided that the source terms reasonably envelop all potential accident scenarios and the robustness of the mitigating features for design basis events can be clearly demonstrated[14].

APPLICATION OF A GRADED APPROACH TO GEOTECHNICAL SITE INVESTIGATION AND CHARACTERIZATION BASED ON RADIOLOGICAL HAZARD CATEGORIZATION

6.8. For nuclear installations categorized as posing a high radiological hazard, the application of a graded approach to the site investigation and characterization is not applicable (see Table 5). The scope of the geotechnical site investigation and characterization should be as described in Sections 2–5.

6.9. The geotechnical site investigation and characterization for installations that do not have on-site or off-site radiological consequences follow the applicable industry standards.

6.10. For nuclear installations categorized as medium or low hazard (see Table 5), the application of a graded approach to the geotechnical site investigation and characterization should be considered. Typically, for an installation in the medium hazard category, a narrower scope than that used for an installation in the high hazard category should be considered. For an installation in the low hazard category, an increased scope compared with that used for an installation in the conventional hazard category should be considered.

6.11. The extent to which a graded approach is applied to the geotechnical site investigation and characterization depends on the foundation needs for the nuclear installation and on the complexity of the subsurface conditions. The appropriate approach should be determined on the basis of the available information and the judgement of qualified geologists and geotechnical and nuclear engineers. A graded geotechnical site investigation and characterization should address, at a minimum, the following items:

(a) The geological structure of subsurface materials, with a description of the stratigraphical sequence of soil or rock strata, and the nature and dimensions in plane and depth of the different formations;

---

[14] The robustness of these features can be clearly demonstrated, for instance, by showing a design margin up to several times the design basis event.

(b)   The static and dynamic geotechnical properties of subsurface materials, as necessary to assess the stability and bearing capacity, to evaluate seismic and other hazards, and to define design basis parameters;

(c)   The potential presence of complex subsurface conditions, such as underground cavities or expansive soils or rocks;

(d)   Hydrogeological conditions[15] at the site, including the presence and thickness of aquifers, the groundwater regime, groundwater levels[16], the amplitude of fluctuations, as well as the chemical composition of groundwater and the potential effects on the materials of underground structures.

The application of a graded approach may include the level of detail (e.g. number and layout of boreholes, types and number of laboratory and field tests) used in the investigation of the items listed above; however, the scope of the geotechnical site investigation should always include these items.[17] Variability and uncertainty in subsurface materials should always be addressed.

GEOTECHNICAL EVALUATION OF SITES FOR NUCLEAR INSTALLATIONS

6.12. Geotechnical site characterization is required in order to provide sufficient information to perform a reliable and defendable site evaluation with respect to geotechnical hazards, including slope instability (see paras 5.27 and 5.28 of SSR-1 [1]), collapse, subsidence or uplift of the site surface (see para. 5.29 of SSR-1 [1]) and soil liquefaction (see paras 5.30 and 5.31 of SSR-1 [1]). A graded approach, depending on site conditions, is required to be applied (see paras 4.4

---

[15] Hydrological parameters important for the characterization of hydrogeological conditions are permeabilities, conductivities, elastic and gravitational water losses, overflows and migration characteristics of aquifers (e.g. distribution coefficient, dispersion). Those parameters are determined by field and laboratory methods and are used to feed groundwater flow models.

[16] The information of interest is usually the time evolution of groundwater levels at different positions around the site, over a long period of time. This information is a key input for the calibration of the groundwater models to be used for the prediction of groundwater conditions at the site.

[17] Defining an appropriate geotechnical site investigation programme for a nuclear installation is site specific, and it is common that the programme is developed in several phases, in which the level of detail is progressively increased, on the basis of the outcome of the previous phase. The application of a graded approach may be achieved by eliminating or reducing the effort in the final phases.

and 4.5 of SSR-1 [1]) and this may mean that simplified bounding analyses or expert judgement could be acceptable to screen out these geotechnical hazards.

6.13. If, as a result of the site evaluation (see Requirement 4 of SSR-1 [1]), one geotechnical hazard cannot be screened out, then a more detailed investigation and characterization should be conducted to refine the evaluation. As a result of this refinement and further evaluation, the site may be considered suitable on the basis of specific established suitability criteria, and corresponding specific design bases should be established to ensure safety throughout the lifetime of the nuclear installation.

DESIGN BASIS OF NUCLEAR INSTALLATIONS DERIVED FROM GEOTECHNICAL SITE CHARACTERIZATION

6.14. The application of a graded approach to the geotechnical site characterization might result in an increased level of uncertainty in the geotechnical parameters used as input for the design basis. This larger uncertainty should be taken into account when defining the design basis.

6.15. The application of a graded approach to the geotechnical site characterization might also result in less detailed knowledge of the structure of the subsurface materials (e.g. variability of soil profiles within the site) or of other characteristics (e.g. physical or geochemical properties of the soil). The design basis should account for such uncertainties by defining reasonable ranges of variation to be considered in the design or by selecting the most unfavourable conditions.

7. APPLICATION OF THE MANAGEMENT SYSTEM TO THE GEOTECHNICAL ASPECTS OF SITES FOR NUCLEAR INSTALLATIONS

7.1. A management system (see Requirement 2 of SSR-1 [1]) applicable to the organizations involved in the geotechnical site investigation, characterization and evaluation should be established before the start of the programme. Requirements for such a management system are established in IAEA Safety Standards Series No. GSR Part 2, Leadership and Management for Safety [12], and supporting recommendations are provided in IAEA Safety Standards Series No. GSG-20, Leadership, Management and Culture for Safety [13].

7.2. The organization responsible for the nuclear installation is required to put in place arrangements with organizations in the supply chain[18] for managing safety (see Requirement 11 of GSR Part 2 [12]). The organizations in the supply chain may have their own management system approved by the main contractor or adhere to the management system of the main contractor. The management system of the main contractor should include arrangements for qualification, selection, evaluation, procurement and oversight of the supply chain.

SCOPE OF THE MANAGEMENT SYSTEM IN RELATION TO THE
GEOTECHNICAL EVALUATION AND MONITORING OF SITES FOR
NUCLEAR INSTALLATIONS

7.3. The management system should cover all the processes and activities described in this Safety Guide, as applicable to each site. This includes the following:

(a) Compilation of data from relevant literature or previous investigations;
(b) Field investigation campaigns, including sampling, logging and storage of samples;
(c) Field testing, measurement or monitoring;
(d) Laboratory testing;
(e) Data processing and reduction of test data;
(f) Calculations and interpretations;
(g) Verification and validation of computer software;
(h) Documentation control and archiving.

---

[18] In the context of this Safety Guide, the supply chain includes site evaluation services, such as area topographical surveying, drilling and sampling, surface geophysics, borehole geophysics, laboratory testing, field testing and field monitoring.

DOCUMENTATION OF THE MANAGEMENT SYSTEM IN RELATION TO THE GEOTECHNICAL EVALUATION OF SITES FOR NUCLEAR INSTALLATIONS

7.4. The documentation describing the management system should be organized into different tiers. The first tier should contain a management system manual including or referring to the following information:

(a)   General statement of policies and objectives of the manual;
(b)   Definition of processes and activities within the scope of the management system;
(c)   Organizational structure for all processes within the scope of the management system, including the responsibility and authority of organizations and personnel involved in the development of those processes;
(d)   Definition of the supervision, review and verification of all processes;
(e)   Description of the planning and conduct of audits and reviews;
(f)   Management of documents, samples and records;
(g)   Provisions for the training of personnel, including the review and verification of training activities;
(h)   List of technical and administrative procedures to be applied, including references to procedures in the second tier of management system documentation.

7.5. Documentation in the second tier should normally be grouped into a manual of management and administrative procedures and a manual of technical procedures.

7.6. Owing to the potentially large variety of investigations and analyses to be performed in relation to geotechnical siting activities, technical procedures and instructions should be developed to facilitate the execution and verification of these activities. These procedures and instructions should normally refer to existing codes and standards, especially for field testing and laboratory testing.[19]

7.7. Each procedure and instruction in the second tier of the management system documentation should include:

(a)   Purpose and scope of the procedure, including prerequisites, precautions and limitations, if applicable.
(b)   Definitions of terms with an uncommon or specific meaning.

---

[19] Many geotechnical correlations or methods use the results of standardized tests; departing from the standardized tests would invalidate these correlations.

(c) References.
(d) Responsibilities, in which the primary responsibility for successful outcome should be identified. The primary responsibility may be different from responsibilities for specific activities.
(e) Qualification and training certifications for personnel.
(f) Actions, or step by step instructions, to be performed to achieve the purpose of the procedure.
(g) Documentation and reports to be produced.
(h) Necessary quality management records and their classification in accordance with the management system manual.

7.8. Procedures should be prepared and reviewed by personnel with sufficient experience in the subject area. These procedures should be evaluated and revised periodically to keep them up to date, as equipment, information, technology, industrial practices and regulatory requirements may evolve.

7.9. To ensure document control, each document should be assigned a unique identification number. Procedures should define how documents are numbered and how obsolete documents are marked to prevent further use.

IMPLEMENTATION OF THE MANAGEMENT SYSTEM IN RELATION TO THE GEOTECHNICAL EVALUATION OF SITES FOR NUCLEAR INSTALLATIONS

Control of studies, evaluations and analyses

7.10. Studies, evaluations and analyses should be peer reviewed by qualified individuals who have not participated in their specification or in their development, with the purpose of ensuring that the intended scope has been met, the technical approach and method of analysis are valid, and the results are correct. In addition, the raw data obtained using recording or measurement instruments should be kept and made available to reviewers, as necessary. Evidence of the review work should be produced and kept as a quality management record in the project archives. The qualifications of the reviewers should be such that they could have competently performed the study, evaluation or analysis they are reviewing.

Control of field activities

7.11. Field activities should be supervised to ensure that they are performed by qualified personnel in accordance with established procedures and using specified equipment. Evidence of this supervision should be produced and kept as a quality management record in the project archives.

Control of samples

7.12. Procedures for the control of samples during handling, storage and shipping should address their cleaning, packing, preservation and identification to prevent the deterioration and loss of samples. The identification of samples of limited lifetime should include the date of acquisition and the expected life.

7.13. The preservation of representative cores from subsurface characterization boreholes may be necessary for a long period of time to allow for additional investigations or interpretations. The period of time during which the cores need to be preserved, as well as the preservation conditions and methods to be used, should be agreed in advance with the regulatory body and specified in the procedures.

Control of laboratory testing

7.14. Specified testing should be performed by accredited laboratories that have been assessed as competent by the organization in charge of the site characterization. Such an assessment is normally based on certificates of qualification issued by an independent organization. Certificates should be kept as quality management records in the project archives.

Control of software

7.15. Commercial software for data acquisition, data processing, evaluations or analyses, used under a licence agreement with the developer, should be installed in accordance with the procedure provided by the developer and checked accordingly. Evidence of this check should be produced and kept as a quality management record in the project archives.

7.16. Commercial software developers should be considered part of the supply chain to the geotechnical site investigation, characterization and evaluation (see para. 7.2). Appropriate certificates issued by the software developers should be kept as quality management records in the project archives.

7.17. For non-commercial software and software developed internally, a verification programme should be developed and run by qualified personnel before the software can be used in the geotechnical site investigation and evaluation. Evidence of the verification work should be produced and kept as a quality management record in the project archives.

7.18. The verification of commercial and non-commercial software does not imply that the mathematical formulation implemented within the software is adequate to represent a particular configuration. The suitability of a piece of software should be assessed on the basis of the available validation information.

Measuring instruments

7.19. The accuracy of measuring instruments should be maintained within prescribed design limits to ensure the necessary reproducibility and traceability of results. Instrument calibration records should be kept as quality management records in the project archives.

7.20. Data processing software used in association with measuring instruments should be verified, as described in paras 7.15–7.18.

Audits, non-conformances and corrective actions

7.21. Periodic audits by a team that is independent from the development team should be performed to verify compliance with the procedures for geotechnical site evaluation and to assess the effectiveness of the management system in order to identify potential improvements.

7.22. The results of audits — including details of non-conformances and the corrective actions derived from them — should be recorded. Reports from audits should be kept as quality management records in the project archives. The implementation of corrective actions should be kept under review, and the closure of non-conformances should be kept as quality management records in the project archives.

7.23. The frequency of audits varies. However, at least one audit should be performed at the project mid-term to ensure that conditions that might adversely affect quality are identified and corrected in time.

APPLYING A GRADED APPROACH TO THE MANAGEMENT
SYSTEM IN RELATION TO THE GEOTECHNICAL EVALUATION OF
SITES FOR NUCLEAR INSTALLATIONS

7.24. The application of a graded approach (see Section 6 of this Safety Guide) to
the management system is also required (see Requirement 7 of GSR Part 2 [12]).
As described in para. 6.10, the application of a graded approach should be
considered for nuclear installations in the medium or low hazard categories.

7.25. The application of a graded approach to geotechnical site evaluation
should involve ensuring that the documentation and administrative effort are
commensurate with the radiological hazard of the installation, while still observing
the main safety objectives, as described in para. 6.3. Further recommendations are
provided in GSG-20 [13]. For example, the application of a graded approach may
result in the following:

(a)   Supplier qualification documentation being accepted without further audits
        or third party certification;
(b)   Review and evaluation being performed on a sample basis;
(c)   The levels of approval of the documentation being reduced;
(d)   Distribution lists being reduced or eliminated;
(e)   The quality records to be generated and retained being reduced;
(f)   The frequency of audits being reduced.

7.26. In whatever way a graded approach is applied, the management system
should, at a minimum, retain the following aspects:

(a)   Definitions of the activities to be performed, with their input, output and
        main guidelines;
(b)   The qualification and training certifications required for personnel;
(c)   The processes for review and evaluation of results;
(d)   Document control.

REFERENCES

[1] INTERNATIONAL ATOMIC ENERGY AGENCY, Site Evaluation for Nuclear Installations, IAEA Safety Standards Series No. SSR-1, IAEA, Vienna (2019).

[2] INTERNATIONAL ATOMIC ENERGY AGENCY, Seismic Hazards in Site Evaluation for Nuclear Installations, IAEA Safety Standards Series No. SSG-9 (Rev. 1), IAEA, Vienna (2022).

[3] INTERNATIONAL ATOMIC ENERGY AGENCY, Seismic Design for Nuclear Installations, IAEA Safety Standards Series No. SSG-67, IAEA, Vienna (2021).

[4] INTERNATIONAL ATOMIC ENERGY AGENCY, Design of Nuclear Installations Against External Events Excluding Earthquakes, IAEA Safety Standards Series No. SSG-68, IAEA, Vienna (2021).

[5] INTERNATIONAL ATOMIC ENERGY AGENCY, Site Survey and Site Selection for Nuclear Installations, IAEA Safety Standards Series No. SSG-35, IAEA, Vienna (2015).

[6] INTERNATIONAL ATOMIC ENERGY AGENCY, Investigation of Site Characteristics and Evaluation of Radiation Risks to the Public and the Environment in Site Evaluation for Nuclear Installations, IAEA Safety Standards Series No. SSG-92, IAEA, Vienna (2025),
https://doi.org/10.61092/iaea.4o4x-cn46

[7] NUCLEAR REGULATORY COMMISSION, Geologic and Geotechnical Site Characterization Investigations for Nuclear Power Plants, RG 1.132, NRC, Washington, DC (2021).

[8] NUCLEAR REGULATORY COMMISSION, A Performance-Based Approach to Define the Site-Specific Earthquake Ground Motion, RG 1.208, Office of Nuclear Regulatory Research, Washington, DC (2007).

[9] INTERNATIONAL ATOMIC ENERGY AGENCY, WORLD METEOROLOGICAL ORGANIZATION, Meteorological and Hydrological Hazards in Site Evaluation for Nuclear Installations, IAEA Safety Standards Series No. SSG-18, IAEA, Vienna (2011). (A revision of this publication is currently in preparation.)

[10] AMERICAN SOCIETY OF CIVIL ENGINEERS, Seismic Analysis of Safety-Related Nuclear Structures, ASCE/SEI 4-16, ASCE, Reston, VA (2016),
https://doi.org/10.1061/9780784413937

[11] INTERNATIONAL ATOMIC ENERGY AGENCY, Design of the Reactor Containment and Associated Systems for Nuclear Power Plants, IAEA Safety Standards Series No. SSG-53, IAEA, Vienna (2019).

[12] INTERNATIONAL ATOMIC ENERGY AGENCY, Leadership and Management for Safety, IAEA Safety Standards Series No. GSR Part 2, IAEA, Vienna (2016),
https://doi.org/10.61092/iaea.cq1k-j5z3

[13] INTERNATIONAL ATOMIC ENERGY AGENCY, Leadership, Management and Culture for Safety, IAEA Safety Standards Series No. GSG-20, IAEA, Vienna (in preparation).

CONTRIBUTORS TO DRAFTING AND REVIEW

Beltran, F.	Belgar Engineering Consultants, Spain
Cetin, K.	Middle East Technical University, Türkiye
Contri, P.	International Atomic Energy Agency
Gulerce, Z.	International Atomic Energy Agency
Hijikata, K.	Tokyo Electric Power Company, Japan
Houston, T.	Consultant, United States of America
Kawai, T.	Tohoku Institute of Technology, Japan
Skytta, P.	University of Turku, Finland
Stoeva, N.	International Atomic Energy Agency
Wang, W.	Nuclear Regulatory Commission, United States of America

Feedback on IAEA publications may be given via the on-line form available at:
www.iaea.org/publications/feedback
The form may also be used to report safety issues or submit environmental queries concerning
IAEA publications.

Alternatively, contact IAEA Publishing directly:

Publishing Section, Dissemination Unit
International Atomic Energy Agency
Vienna International Centre, PO Box 100, 1400 Vienna, Austria
Telephone: +43 1 2600 22529 or 22530
Email: sales.publications@iaea.org
www.iaea.org/publications

ORDERING LOCALLY

To purchase priced IAEA publications, please contact either your preferred local supplier
or the IAEA's lead distributor:

Mare Nostrum Group
39 East Parade
Harrogate
North Yorkshire
HG1 5LQ
United Kingdom
Email: enquiries@mare-nostrum.co.uk
www.mngbookshop.co.uk

Trade Orders and Enquiries:
Telephone: +44 1243 843 291
Email: trade@wiley.com

Individual Orders and Enquiries:
Email: mng.csd@wiley.com

Priced and unpriced IAEA publications may also be ordered directly from the IAEA by contacting
IAEA Publishing. The recipient is responsible for shipping costs and any customs and duties.

Printed and bound by CPI Group (UK) Ltd, Croydon, CR0 4YY

06/07/2026

02160597-0003